To Build New York

100 YEARS OF INFRASTRUCTURE

Other Books in the McGraw-Hill Construction Series

Defect-Free Buildings: A Construction Manual for Quality Control and Conflict Resolution by Robert S. Mann

Building Anatomy: An Illustrated Guide to How Structures Work by Iver Wahl

Construction Safety Engineering Principles: Designing and Managing Safer Job Sites by David V. MacCollum

McGraw-Hill Construction Locator: Building Codes, Construction Standards, Project Specifications, and Government Regulations by Joseph A. McDonald

Building Information Modeling: Planning and Managing Construction Projects with 4D CAD and Simulations by Willem Kymmell

Urban Construction Project Management by Richard Lambeck and John Eschemuller

Solar Power in Building Design: The Engineer's Complete Design Resource by Peter Gevorkian

The Engineering Guide to LEED-New Construction: Sustainable Construction for Engineers by Liv Haselbach

Emerald Architecture: Case Studies in Green Building by GreenSource: The Magazine of Sustainable Design

Green Building Through Integrated Design by Jerry Yudelson

The Green Building Bottom Line: The Real Cost of Sustainable Building by Martin Melaver and Phyllis Mueller

About McGraw-Hill Construction

McGraw-Hill Construction, part of The McGraw-Hill Companies (NYSE: MHP), connects people, projects and products across the design and construction industry. Backed by the power of Dodge, Sweets, Engineering News-Record (ENR), Architectural Record, GreenSource, Constructor, and Regional Publications, the company provides information, intelligence, tools, applications and resources to help customers grow their business. McGraw-Hill Construction serves more than one million customers within the $4.6 trillion global construction community. For more information, visit www.construction.com.

Columbus Circle by day, early 20th century. Courtesy New York Transit Museum.

Columbus Circle at night. Courtesy Tully Construction Co. Inc.

George Washington Bridge by day, early 20th century. Courtesy Port Authority of New York and New Jersey.

George Washington Bridge at night. © Bettmann/CORBIS.

Constructing New York City's first water tunnel. Courtesy New York City Department of Environmental Protection.

A man works in an underground water valve chamber in the Bronx, New York. Courtesy Getty Images.

To Build New York

100 YEARS OF INFRASTRUCTURE

The General Contractors Association of New York

Mc
Graw
Hill

New York Chicago San Francisco Lisbon London Madrid
Mexico City Milan New Delhi San Juan Seoul
Singapore Sydney Toronto

Library of Congress Cataloging-in-Publication Data

To build New York : 100 years of infrastructure /The General Contractors Association of New York. -- 1st ed.
   p. cm.
Includes index.
ISBN 978-0-07-160862-6 (alk. paper)
1. Public works--New York (State)--New York--History. 2. Municipal engineering--New York (State)--New York--History. 1. General Contractors Association of New York.
TA25.N72T62 2009
363.6097471--dc22

200802525059

To Build New York: 100 Years of Infrastructure

1 2 3 4 5 6 7 8 9 0 DOW/DOW 0 1 2 1 0 9 8

ISBN 978- 0-07-160862-6
MHID 0-07-160862-1

Sponsoring Editor	**Editorial Director**	**Composition**
Joy Bramble Oehlkers	Felice Farber	Carmen Group Inc.
Acquisitions Coordinator	**Proofreader**	**Art Director, Cover**
Rebecca Behrens	Bea Ruberto	Renee Klein
Editorial Supervisor	**Indexer**	**Historian**
David E. Fogarty	WordCo Indexing Services	Benjamin Miler
Copy Editor	**Production Supervisor**	**Writer**
Roberta Mantus	Richard C. Ruzyzka	David Kusner

Contents

Foreword	xvii
Introduction	1
Bridges	6
Water & Sewers	8
Mass Transit	10
Roads	12
TIMELINE: 1909–1929	**14**
1909–1929: BUILDING A GREATER NEW YORK	**16**
Digging: Going Down	34
Tunneling: Going Sideways	41
TIMELINE: 1930–1945	**48**
1930–1945: IN DIFFICULT TIMES, BUILDING THE FOUNDATION FOR THE MODERN METROPOLIS	
Foundations: Pushing Up	50
"Dewatering" in an Island City	62
	65
TIMELINE: 1946–1962	**70**
1946–1962: CAPITAL OF THE WORLD	**73**
Underpinning	88
TIMELINE: 1963–1979	**92**
1963–1979: URBAN CRISIS	**95**
Concrete	108
TIMELINE: 1980–2009	**112**
1980–2009: REVIVAL	**114**
Asphalt Paving	130
GCA Members	134
GCA Presidents	152
Photo Credits	158
Notes	160
GCA Memorabilia	164
Index	167

★ ★ ★

THE GENERAL CONTRACTORS ASSOCIATION REPRESENTS

THE BUILDERS OF NEW YORK'S HEAVY INFRASTRUCTURE.

IT WAS FORMED IN 1909.

IN 2009 IT WILL CELEBRATE ITS CENTENNIAL.

100 YEARS OF INFRASTRUCTURE:

1909-2009

Aerial view of Triborough Bridge with Hell Gate Bridge and Wards Island. Wastewater Treatment Plant in the foreground. Photographer: Roger Swingle. www.swingleprints.com

Foreword

When I took the oath of office on the steps of City Hall, just more than 100 days after the terrorist attacks of September 11th, I could see smoke still rising from the World Trade Center site. The attacks had killed 2,751 people and plunged our city into mourning. Yet at the worst moment in our history, New Yorkers were at their best — selfless, courageous, heroic. More than 400 firefighters, police officers, and emergency service workers gave their lives to save the lives of more than 25,000 of their fellow citizens who were able to escape from the doomed office buildings and the adjoining area in a safe and orderly way.

New York City's construction contractors and construction workers also answered the call, helping to rescue survivors, recover bodies, and begin to clear the area that came to be called "Ground Zero." Months later, ahead of schedule, the area had been cleared of debris, and civic, government, and business leaders began to develop plans for renewing Lower Manhattan and constructing an inspiring memorial.

As we developed plans to rebuild, we marveled at how well those who came before us had built the physical structures that make our great metropolis possible. In spite of the collapse of our two tallest skyscrapers, the subway lines, water mains, sewers, electric lines, and gas and steam pipes did not fail. Soon enough, Lower Manhattan was open for business, and as a result of the investments we have made in parks, housing, schools, and other infrastructure, it has become one of the fastest growing residential communities in the country.

New York City's post-9/11 comeback has been an inspiring story, but it is only the latest example of New Yorkers responding to crisis — and rising to the challenge of building an even greater city. *To Build New York: 100 Years of Infrastructure* tells the story of how previous generations built and rebuilt the world's greatest metropolis. In the face of wars, depression, recessions, fiscal crises, and social conflicts, New Yorkers constructed the critical infrastructures — bridges, tunnels, subways, highways, water mains, airports, piers, power lines, and sewage treatment plants, among other facilities — that knit together five boroughs into one great city that became the world's economic and cultural capital.

The book's fascinating photographs give life to the enormity, complexity, and difficulty of the city's great construction projects — and the backbreaking work involved. When we get on the subway or turn on the water or flip a switch in our homes, we can think of the faces we see in this book.

In a sense, *To Build New York* is the challenge that every generation faces and meets. Over the decades, New York City has always had to maintain and repair existing structures and envision and construct new infrastructure in order to accommodate residents, attract businesses, make use of new technologies, and meet the challenges of an increasingly competitive global marketplace. As New Yorkers continue the journey that brought us from the era of the trolley car to the age of the Internet, three common elements stand out from every period of our history:

New Yorkers are resilient. Our city has survived the Great Fire, the Great Flu, the Great Depression, and 9/11. And we have emerged from each of the crises — and many more — stronger than ever.

New Yorkers are visionaries. We have always been a city that continually reimagines, reinvents, and rebuilds itself. We recognize that physical structures are essential for a city to sustain vibrant communities and a high quality of life. City agencies, such as New York City's Landmarks Preservation Commission, have been at the forefront to assure that we preserve our historically significant buildings and neighborhoods while the New York City Planning Commission, the New York City Design Commission, and the Department of Design and Construction all work to assure that our city's structures are designed and built to meet the highest standards. And PlaNYC has emerged as a national leader in setting the standard for building environmentally sustainable cities in the twenty-first century.

New Yorkers are known for their ambition and for their capacity to work together, to harness the energy of business, labor, and government, in order to build the great projects that define our city. Our municipal workforce has been an integral element in the team of construction firms and workers that have created our magnificent parks, museums, roadways, and water system.

New Yorkers just don't quit. In every era, we have been through tough times. And always we came out stronger because we came together to make the city better for everyone. At every decisive moment, New Yorkers have always made the hard choices, painful sacrifices, and bold investments to build a better future.

This wonderful book tells the inspiring stories of the dreamers and doers who built the city that never quits, never rests on its laurels, and never loses faith that tomorrow will be better than today. Best of all, it lets today's readers take a look at the construction contractors, municipal and construction workers, and construction projects that created New York City as we know it. Enjoy the photos. Learn from the history. Appreciate the anecdotes. Honor those who have worked so hard to construct the physical structures that make our metropolis such a great place to live and work and visit.

And then, take up the challenge that defines every generation's life and legacy — To Build New York, greater than ever.

Michael R. Bloomberg
Mayor

Queens Midtown Tunnel Holing Through Ceremony
Mayor Fiorello LaGuardia pulling a switch, setting off explosives to clear the remaining six foot wall of rock between the Manhattan and Queens sides of North and South Tunnels. Time: 11:44 AM. November 8, 1939. Photoflash. Photographer: Voss Studios. Courtesy MTA Bridges and Tunnels Special Archive.

Interborough Rapid Transit (IRT)
Aerial view of City Hall during the opening day ceremonies of the Interborough Rapid Transit (IRT), November 27, 1904. Courtesy New York Transit Museum.

Opening, Flushing Bridge, April 26, 1939
Courtesy LaGuardia and Wagner Archives, LaGuardia Community College/The City University of New York.

Introduction

New York City is not only the world's leading metropolis. It is a modern day miracle.

Located on three islands and one tip of the mainland, New York City is joined together by highways, bridges, tunnels, and subways. Every day, more than 8 million people—about the same number as the city's population—ride the subways. About 800,000 cars and trucks enter and exit the city, 1.3 billion gallons of water pour through 6,000 miles of plumbing, and 50 million megawatts of electricity course through the power grid. Rail lines, three airports, and seaports connect the city with the rest of the country and every continent.

Without this physical infrastructure, New York would not have been able to forge five boroughs into one city, welcome millions of newcomers, and become the financial and cultural capital of the world. When the modern municipality of New York City was founded in 1898, traveling from the outskirts of the outer boroughs to the offices, factories, and stores in Manhattan was difficult and time-consuming. Horse-drawn vehicles and trolley cars clogged the streets, while the elevated rail lines were confined to Manhattan and were not connected to each other. Water quality was poor, poor people suffered from diseases such as yellow fever, and the wealthy often fled the city for the sake of their health.

Coming soon, another project "built by Moses."
Courtesy LaGuardia and Wagner Archives, LaGuardia Community College/The City University of New York.

Building this great metropolis—its roads and highways, bridges and tunnels, airports and aqueducts, subways, skyscrapers, and sewage treatment plants—required a partnership with New York City government and a cooperation among construction contractors, craft workers and their unions, the business community, the state and federal governments, and other sectors of society from academia to charitable institutions. Founded in 1909, the General Contractors Association of New York (GCA) has brought together the city's public works general contractors to do our part to build the metropolis.

Over the past 100 years, the GCA has participated in the epochal effort—involving ingenuity, skill, risk, and often enormous costs in public and private investments and, all too often, life and limb as well—that built the world's most complex and effective infrastructure. When New York City's residents, commuters, and visitors drive on our roadways, ride on our subways, live and work in our towering buildings, drink the water, and leave or arrive by way of our railroads or airports, they may marvel at the feats of engineering that they represent. But they seldom ask about the contractors and construction workers who built them.

This is the purpose of this book. To tell who built what elements of New York's infrastructure, when, and where. And, perhaps even more importantly, to give a sense of how and why.

1

Opening the Steinway Tunnel to traffic, June 22, 1915. Courtesy General Contractors Association of New York.

Lexington Ave/53rd Street ribbon cutting ceremony with John Gilhooley and Daniel Scannel in early 1960s. Courtesy New York Transit Museum.

Preinspection tour. Courtesy New York Transit Museum.

125th Street and Broadway IRT. Preopening subway inspection trip. Courtesy New York Transit Museum.

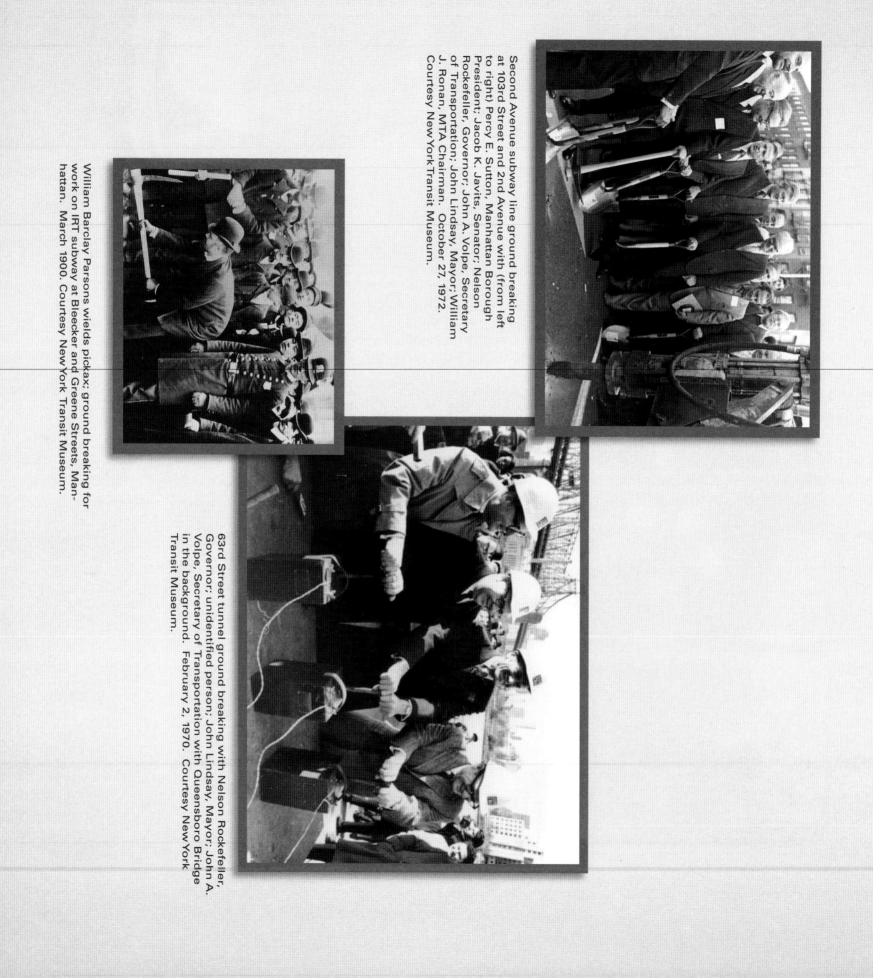

Second Avenue subway line ground breaking at 103rd Street and 2nd Avenue with (from left to right) Percy E. Sutton, Manhattan Borough President; Jacob K. Javits, Senator; Nelson Rockefeller, Governor; John A. Volpe, Secretary of Transportation; John Lindsay, Mayor; William J. Ronan, MTA Chairman. October 27, 1972. Courtesy New York Transit Museum.

William Barclay Parsons wields pickax; ground breaking for work on IRT subway at Bleecker and Greene Streets, Manhattan. March 1900. Courtesy New York Transit Museum.

63rd Street tunnel ground breaking with Nelson Rockefeller, Governor; unidentified person; John Lindsay, Mayor; John A. Volpe, Secretary of Transportation with Queensboro Bridge in the background. February 2, 1970. Courtesy New York Transit Museum.

Inspection by Mayor LaGuardia and others of City Water Tunnel 2. Queens Division, Section 1 Contract 226. Shaft 9A. GCA member Patrick McGovern Inc., contractor. January 6, 1936. © New York City Department of Environmental Protection.

Ground breaking for the construction of the Holland Tunnel. Pictured second from right is Clifford Milburn Holland. October 12, 1920. Courtesy Port Authority of New York and New Jersey.

Connecting New York

Bridges

1909   Queensboro Bridge
1909   Manhattan Bridge
1910   Madison Avenue Bridge
1910   Hunters Point Avenue Bridge
1917   Hell Gate Bridge
1917   Ocean Avenue Bridge
1922   Eastchester Bridge
1923   Hook Creek Canal Bridge
1925   North Channel Bridge
1925   Nolins Avenue Bridge
1927   Hawtree Basin Bridge
1927   Roosevelt Avenue Bridge
1928   Goethals Bridge
1928   Outerbridge Crossing Bridge
1928   174th Street Bridge
1929   Greenpoint Avenue Bridge
1929   Stillwell Avenue Bridge
1931   George Washington Bridge
1931   Bayonne Bridge
1931   Hook Creek Bridge
1931   Little Neck Bridge
1931   Eastern Boulevard Bridge
1931   Fresh Kills Bridge
1931   Cropsey Avenue Bridge
1932   Metropolitan Avenue Bridge
1932   238th Street Bridge
1933   Harway Avenue Bridge
1936   Triborough Bridge
1936   Eastern Boulevard Bridge
1937   Henry Hudson Bridge
1938   Marine Parkway Bridge
1938   Westchester Avenue Bridge
1939   Cross-Bay/Veterans' Memorial Bridge
1939   Flushing/Northern Boulevard Bridge

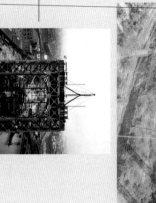

Kosciuszko Bridge 1939
Bronx-Whitestone Bridge 1939
Whitestone Expressway Bridge 1939
Mill Basin Bridge 1940
Midtown Highway Crossing Bridge 1940
Hutchinson River Parkway Bridge 1941
Hamilton Avenue Bridge 1942
Wards Island Bridge 1951
Unionport Bridge 1953
Bruckner Boulevard/Ecstern Boulevard Bridge 1953
Pulaski/Vernon Avenue Bridge 1954
New York Central Harlem River Bridge 1954
Roosevelt Island Bridge 1955
Park Avenue Bridge 1956
Lemon Creek Bridge 1958
Arthur Kill Railroad Lift Bridge 1959
Throgs Neck Bridge 1961
Broadway Bridge 1962
"Martha Washington"/GWB Lower Level Bridge 1962
Hawtree Basin Bridge 1963
Alexander Hamilton Bridge 1963
Verrazano-Narrows Bridge 1964
Rikers Island Bridge 1966
Eastchester Bridge 1966
Cross Bay Bridge 1970

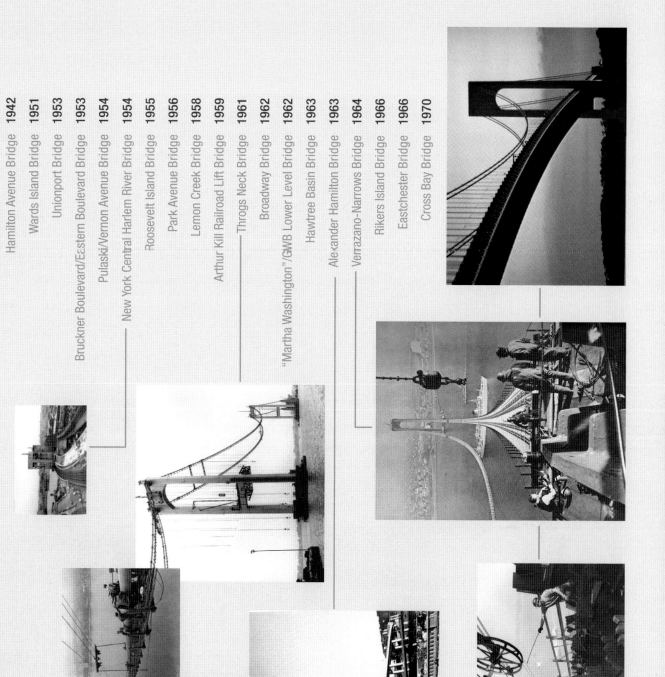

Connecting New York

Water & Sewers

1909 Croton Falls & Croton Falls Diverting Reservoir
1909 Ashokan Reservoir
1909 Catskill Aqueduct System
1911 Steel pipe siphon laid across Narrows to Staten Island
1911 City Water Tunnel No. 1 delivers first water to Manhattan
1915 Gowanus Flushing Tunnel
1915 Catskill Aqueduct delivers first water to Kensico Reservoir
1917 Shandaken Tunnel
1919 Schoharie Reservoir
1927 Water Tunnel No. 2
1931 US Supreme Court allows NYC to use Delaware River water; work begins on Delaware Aqueduct
1935 Coney Island Wastewater Treatment Plant
1935 Bronx Wards Island Interceptor Sewer
1935 Bowery Bay Wastewater Treatment Plant
1937 Rondout Reservoir
1937 Jerome Avenue Interceptor Sewer
1937 Wards Island Wastewater Treatment Plant
1939 Tallman's Island Wastewater Treatment Plant
1940 City Island Hart Island Wastewater Treatment Plant
1941 Neversink Reservoir
1943 Jamaica Wastewater Treatment Plant
1944 26th Ward Wastewater Treatment Plant
1945 Neversink construction resumes
1945 Rondout construction resumes
1947 Whitestone Interceptor Sewer
1947 Pepacton Reservoir
1949 East Bronx Interceptor Sewer
1949 Hunt's Point Wastewater Treatment Plant
1949 Owl's Head Wastewater Treatment Plant
1950 Chelsea Pump Station
1951 Port Richmond Wastewater Treatment Plant

Cannonsville Reservoir **1955**
Oakwood Beach Interceptor Sewer **1955**
Rector Street Interceptor Sewer **1955**
Johnson Avenue Interceptor Sewer **1957**
Kent Avenue Interceptor Sewer **1958**
Morgan Avenue Interceptor Sewer **1960**
Greenpoint Avenue Interceptor Sewer **1960**
Richmond Tunnel **1964**
South Branch Interceptor Sewer **1965**
Silver Lake Storage Tanks **1967**
Newtown Creek Interceptor Sewer **1967**
West Side Interceptor Sewer **1967**
Water Tunnel No. 3 **1969**
West Branch Interceptor Sewer **1969**
West Street Interceptor Sewer **1971**
North River Wastewater Treatment Plant **1972**
Work halted because of fiscal crisis **1974**
Bay Street Interceptor Sewer **1974**
Richmond Terrace Interceptor Sewer **1975**
Eltingville Interceptor Sewer **1977**
Red Hook Interceptor Sewer **1978**
Water Tunnel No. 3 restarts **1979**
DEP begins replacing 6-inch water mains **1980**
(which are more fragile than larger mains;
the work continues to the present)
Water Tunnel No. 3 Stage 1 complete **1983**
Flushing Water Quality Service **1985**
Red Hook Wastewater Treatment Plant **1987**
Water Tunnel Stage 1 into service **1998**
Water Tunnel Stage 2 in construction **2001**
Croton Filtration Plant construction begins **2005**

Connecting New York

Mass Transit

1909	IRT construction continuing
1913	Centre Street Loop
1913	Williamsburg Bridge to Chambers Street
1915	IRT Steinway Tunnels open
1915	Flushing Line
1915	Grand Central to Queensboro Place
1915	Sea Beach Line
1915	Myrtle Avenue Extension
1915	Fullton Street Extension
1916	Fourth Avenue to 86th Street
1917	Flushing Line
1917	Queensboro Place to 103rd Street
1917	Astoria Line IRT
1917	West End Line
1922	Eastern Parkway Line to New Lots Avenue
1924	14th Street–Eastern District Line
1924	Sixth Avenue to Montrose Avenue

Fourth Avenue Line to 95th Street   **1925**
Flushing Line to Times Square   **1927**
Flushing Line to Main Street   **1928**
14th Street–Eastern District Line to Broadway Junction   **1928**
Centre Street Loop   **1931**
Chambers Street to Battery   **1931**
14th Street–Eastern District Line to Eighth Avenue   **1931**
IND Eighth Avenue opens   **1932**
IND Sixth Avenue opens   **1940**
63rd Street East River Tunnel starts   **1969**
Second Avenue Tunnel starts   **1972**
Second Avenue Tunnel stops   **1975**
63rd Street East River Tunnel completed   **1989**
IND Queens Boulevard Line East River Tunnel   **1998**
7 Train Extension   **2007**
Second Avenue Tunnel restarts   **2008**

Connecting New York

Roads and Tunnels

1915 Bronx River Parkway
1927 Holland Tunnel
1929 Miller Highway
1929 West Side Highway
1929 Joe DiMaggio Highway
1931 Grand Central Parkway
1931 Northern State Parkway
1933 Interboro Parkway/Jackie Robinson Parkway
1933 Henry Hudson Parkway
1934 Belt Parkway
1934 Cross Island Parkway
1935 Franklin D. Roosevelt Drive
1935 Pelham Parkway
1935 Mosholu Parkway
1936 Whitestone Expressway
1936 Hutchinson River Parkway (Bronx Portion)
1937 Brooklyn-Queens Expressway
1937 Lincoln Tunnel
1939 Long Island Expressway
1939 Gowanus Parkway
1940 Queens Midtown Tunnel

Harlem River Drive **1947**
Van Wyck Expressway **1947**
Cross-Bronx Expressway **1948**
Major Deegan Expressway **1949**
Brooklyn Battery Tunnel **1950**
New England Thruway (Bronx Section) **1950**
Prospect Expressway **1953**
East River Drive **1953**
Sunrise Highway **1953**
Horace Harding Expressway/Long Island Expressway **1954**
Queens Midtown Expressway **1955**
Bruckner Expressway **1957**
Clearview Expressway **1957**
Throgs Neck Expressway **1958**
Sheridan Expressway **1958**
Cloves Lake Expressway/Staten Island Expressway **1959**
West Side Expressway **1959**
Willowbrook Expressway/Dr. Martin Luther King Jr. **1962**
Richmond Parkway/Korean War Veterans Parkway **1966**
Nassau Expressway **1967**

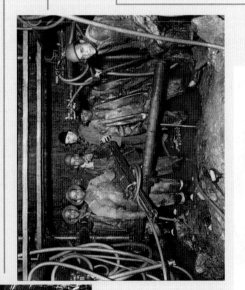

Timeline: 1909–1929

14

1909
- Queensboro Bridge opens
- Manhattan Bridge opens
- GCA founded
- Mayor George Brinton McClellan, Jr.
- Subway underway

1910
- Madison Avenue Bridge
- Hunters Point Avenue Bridge
- Mayor William J. Gaynor
- Chelsea Piers completed
- Pennsylvania Station, tunnels open

1911
- Gowanus Flushing Tunnel
- West 138th Street
- Croton Falls and Croton Falls Diverting Reservoir

1912
- Ashokan Reservoir

1913
- Woolworth Building
- Grand Central Station
- Centre Street Loop Williamsburg Bridge to Chambers Street
- Astoria Gas Plant, Con Edison, in operation

1914
- Mayor John Purroy Mitchel
- Flushing Line, Grand Central to Queensboro Plaza
- Sea Beach Line
- Myrtle Ave. Extension
- Fulton St. (Liberty Ave.) Extension
- IRT Steinway Tunnels
- West 206 Marine Transfer Station
- Catskill Aqueduct delivers first water to Kensico Reservoir

1915
- NYS Barge Canal
- Fourth Ave. Line to 86th St.
- Steel pipe siphon laid across Narrows to Staten Island
- Ocean Avenue Bridge (pedestrian)
- Flushing Line, Queensboro Plaza to 103rd St
- City Water Tunnel No.1 delivers first water to Manhattan

1916
- New York Connecting Railroad
- US joins WWI
- Hell Gate Bridge
- Dual Contract
- Seventh Avenue Line
- Lexington Avenue Line
- Astoria Line IRT
- West End Line

1917
- Last horsedrawn carriage
- Brooklyn Army Terminal
- Lexington "H" System
- Queensboro Plaza to Lexington Avenue Line
- Broadway Line connection to DeKalb Ave. Station via Manhattan Bridge
- Astoria Line BRT
- Jamaica Line

1918
- WWI ends
- Mayor John F. Hylan
- Jerome Avenue Line
- Broadway Line Whitehall St. to Times Square

TIMELINE: 1909-1929

1919
- Seventh Avenue Line connection to Brooklyn
- Culver Line

1920
- Jamaica Bay Pelham Line
- White Plains Road Line
- Eastern Parkway Line to Utica Ave.
- Nostrand Avenue
- Broadway Line to Queensboro Plaza
- Broadway Line connection to DeKalb Ave. Station via tunnel
- Brighton Line

1921
- Port Authority created

1922
- Eastchester Bridge
- Staten Island Piers
- Eastern Parkway Line to New Lots Ave.

1923
- Hook Creek Canal Bridge
- Tiffany Street Marine Transfer Station

1924
- 14th Street-Eastern District Line Sixth Avenue to Montrose Ave.
- Shandaken Tunnel

1925
- North Channel Bridge
- Nolins Avenue Bridge
- Bronx River Parkway
- Fourth Ave. Line to 95th St.
- Hudson Avenue Plant (Con Ed) completed sometime "mid-twenties"
- Hell Gate Plant, Con Ed, in construction

1926
- Mayor James J. Walker
- Hunts Point Gas Plant, Con Ed
- 14th Street and East River (Con Ed) plant opens
- Schoharie Reservoir

1927
- Hawtree Basin Bridge
- Roosevelt Avenue Bridge
- Flushing Line to Times Square
- Holland Tunnel opens
- Catskill Aqueduct system completed

1928
- Goethals Bridge opens
- Outerbridge Crossing opens
- 174th Street Bridge
- Flushing Line to Main St.
- 14th Street-Eastern District Line to Broadway Junction
- East 73rd Street Marine Transfer Station
- Brooklyn Union Gas completes Greenpoint/Newtown Creek plant

1929
- Greenpoint Avenue Bridge
- Stillwell Avenue Bridge
- NYS imposes a gasoline tax which funds road and bridge construction
- Depression
- Regional Plan Association releases its first plan

1909 – 1929:
Building a *Greater New York*

In 1609, an English explorer named Henry Hudson, who had been hired by the Dutch to find a faster route to China, sailed into the waters that are now called New York Harbor. There he saw the island that was soon known as Manhattan. Hudson never dreamed that the river to the west of the island would be named after him. Nor could he ever have imagined that he would be remembered as a founder of what became the world's greatest city.

Three centuries later, New York City celebrated its 300th birthday with a festival that featured a million electric lights on its bridges and buildings, a long procession of naval ships through its harbor and rivers, and a solitary airplane, piloted by the pioneering aviator Wilbur Wright, flying within 25 feet of the Statue of Liberty. Called the Hudson-Fulton Celebration, this festival was a testament to New Yorkers' pride in their past and faith in their future.[1]

New Yorkers had much to be proud of, much to look forward to, and much cause for concern. Only 11 years earlier, the modern municipality of New York City had been formally created. Joining together with Manhattan, which already called itself New York City, were the separate city of Brooklyn and the bedroom communities of the Bronx, Queens, and Staten Island. Together, these five "boroughs" comprised what was far and away the

largest city in the United States. As the lawyer, city planner, and civic leader Andrew Haswell Green declared, the consolidated city immediately became a commercial, cultural, and financial capital that rivaled London for leadership of the world.

New York City's future seemed limitless, but so did its challenges. Towering over the harbor, the Statue of Liberty welcomed the world's peoples with these famous words: "Give me your tired, your poor, your huddled masses yearning to breathe free." From 1882 through 1922, more than 16 million immigrants came to the United States through New York City, and about 3 million remained in the metropolis. Not far from the mansions of the wealthy and the homes of the established middle and working classes were those whom the journalist Jacob Riis described as "the other half." They were mostly recent immigrants and their children who lived in slums, worked in sweatshops, and struggled to build better lives for their children in their new country. In 1909, the discontent of many of these poverty-stricken workers spilled over into the streets, when the city's garment workers went on strike in what was called "the great uprising of the 20,000."

The great metropolis's great challenges were reflected not only in its socioeconomic inequalities but also in its physical infrastructures.

Grand Central Station
(opposite page)

© Photo Collection Alexander Alland, Sr./ CORBIS. Photographer Robert L. Bracklow. Manhattan, New York, New York

IRT Corona Line (above)

Queens Boulevard, Queens. When an elevated line was to cross or traverse an important boulevard, special attention was paid to the appearance of these structures. Such was the case of the elevated line that ran above Queens Blvd. This design called for an ornamental structure of reinforced concrete. November 25, 1913. Courtesy New York Transit Museum.

"What am I proudest of? We played a significant role in the redevelopment of Grand Central Terminal. We had a series of contracts with Metro North and did projects over there for 10 to 12 years from the mid-80s through 1997."

James Moriarty, Jr., *President, T. Moriarty and Son*
President, GCANY

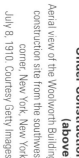

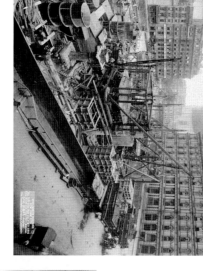

The Woolworth Building Under Construction (above)

Aerial view of the Woolworth Building construction site from the southwest corner, New York, New York, July 8, 1910. Courtesy Getty Images.

Queensboro Bridge (right)

Queensboro Bridge under construction. Courtesy General Contractors Association of New York.

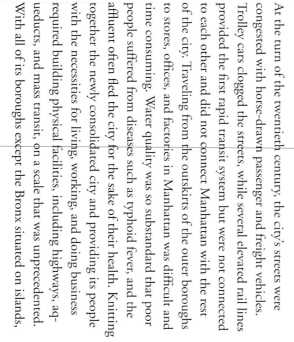

REMARKS

Length of Dam Proper   1168 Feet

At the turn of the twentieth century, the city's streets were congested with horse-drawn passenger and freight vehicles. Trolley cars clogged the streets, while several elevated rail lines provided the first rapid transit system but were not connected to each other and did not connect Manhattan with the rest of the city. Traveling from the outskirts of the outer boroughs to stores, offices, and factories in Manhattan was difficult and time consuming. Water quality was so substandard that poor people suffered from diseases such as typhoid fever, and the affluent often fled the city for the sake of their health. Knitting together the newly consolidated city and providing its people with the necessities for living, working, and doing business required building physical facilities, including highways, aqueducts, and mass transit, on a scale that was unprecedented. With all of its boroughs except the Bronx situated on islands,

the city also needed new physical structures to transport people and products to and from the rest of the country. Moreover, only by improving its port facilities could New York fulfill its potential as a center of commerce for the entire world.

Little more than a decade after its five boroughs joined together, New York City was developing the vision, devising the plans, and beginning the work of building a physical infrastructure on a scale comparable to the metropolis' extraordinary concentration of humanity, commercial activity, and creative energy. New York City had a huge job to do, and it required remarkable teamwork by the city's government, businesses, construction contractors, and construction workers to get the job done.

Guiding the Growth

By the early twentieth century, the use of steel beam construction and improved elevators made it possible to build much taller office and apartment buildings. The first skyscrapers, such as the 42-story Equitable Building, sprouted up, and New York's world-renowned skyline began to emerge.

Growing numbers of New Yorkers expressed concerns about the loss of light and air, as well as housing shortages and the encroachments of factories and warehouses upon residential areas. In response to demands for laws regulating land use, in 1916, New York City adopted its first zoning resolution. It set limits on how tall buildings could be built, how far from the street they needed to be, and whether different sections of the city would be reserved for residential, commercial, or industrial uses. The new zoning ordinance set the tone for construction in New York City, encouraging tall, narrow office buildings and three- to six-story apartment buildings.[2]

In 1909, the same historic year that New York City staged its tricentennial celebration, a group of construction companies banded together to help public agencies and private industry build these bridges, highways, aqueducts, subway systems, office buildings, sewage treatment plants, and other essential projects. Over the 10 ensuing decades—New York City's fourth century—the General Contractors Association did its part to build New York City. In the process, GCA has contributed to the city's economic, social, and cultural progress. GCA members have built, overhauled, and maintained the huge and heavy physical structures that are indispensable to the world's greatest metropolis. GCA members have helped to build the middle class in the New York City metropolitan area and throughout the nation by providing generations of construction workers with good paying, unionized jobs, with portable health and pension programs and joint labor management apprenticeship and training programs in crafts, in new technologies, and in job safety. GCA members have set the standard for quality work, and have changed the way that construction is done throughout the world, leading the way in new technologies and protection for the health and safety of construction workers and the general public. In times of crisis, from the Great Depression of the 1930s to the fiscal crisis of the 1970s and the terrorist attacks of September 11, 2001, GCA members have helped New York City to rebuild its infrastructure and reclaim its future.

Building New York has been an unprecedented and often unheralded accomplishment. This book tells the story, beginning with the extraordinary accomplishments of the first two decades after the GCA's founding. During these decades, New York City built its skyscrapers; expanded its water supply; built new bridges, tunnels, and rail lines and links; and developed the visionary plans and the powerful public agencies that would enable the city to continue growing and prospering in the years ahead.

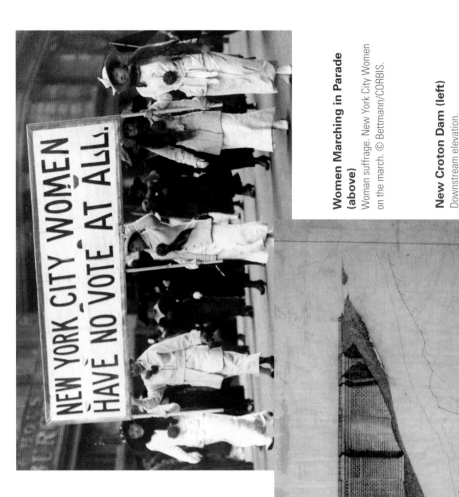

Women Marching in Parade (above)
Woman suffrage. New York City Women on the march. © Bettmann/CORBIS.

New Croton Dam (left)
Downstream elevation.
© NYC Department of Environmental Protection.

"The People Ride in a Hole in the Ground"

As early as 1888, New York Mayor Abram Hewitt called for connecting the people and commerce of Manhattan, Brooklyn, Queens, and the Bronx with a vast system of underground railways.[3] While London, Glasgow, Budapest, and Boston had all built such systems, New York City's subway system would soon become the world's largest. Construction began in 1900 and would continue throughout the twentieth century, becoming one of the city's most ambitious and important endeavors.

By 1900, the city had issued a contract for construction, and work had begun in earnest. The project was designed by the legendary engineer William Barclay Parsons of the city's Rapid Transit Commission and was built and operated by a private venture, the Interborough Rapid Transit Company (IRT), headed by the wealthy financier August Belmont, one of the richest men in New York. Under the direction of several general contractors (including GCA members the Degnon-McLean Construction Company; the John C. Rodgers Company; and the Sicilian Asphalt Paving Company[4]), the construction workers trenched and blasted their way through 19 miles of streets and bedrock in only four years, beginning what would be the largest, fastest, and most crowded subway system in the world. Accomplishing this huge and heroic task were 7,700 workers, mostly Irish and Italian immigrants, who were mostly paid about $2 a day.[5] The new line—soon called the IRT—stretched from the lower tip of Manhattan to the Upper West Side.

From its opening on October 27, 1904, the new subway was an immediate success. In fact, it was so successful that it soon was overcrowded, creating the demand that more subway lines be built. From the first day, there were not nearly enough seats for every passenger, so many riders stood in the middle of the cars, hanging onto leather straps attached to the tops of the trains. "God bless the straphanger," Belmont said, coining a new term for the city's hurried, harried, but basically good-natured commuters.

Soon, most New Yorkers agreed that more subways had to be built, but there was no consensus about who would build them or who would pay for them. Not Belmont nor his competitors nor the city government would provide the money to build more subways all by themselves.

Eventually, a complex compromise was reached. In addition to the IRT, another company, the Brooklyn-Manhattan Transit Company (soon known as the BMT) would also operate subway lines, while a team of contractors would build the new lines. A group of construction companies, including four GCA members—Patrick J. McGovern,[6] the Degnon Company; Underpinning and Foundation, and the Bradley Contracting Company[7]—were put to work. Soon they were digging, boring, and blasting their way through everything from asphalt to quicksand and bedrock. Often they worked under the elevated lines. Within several years, the contractors and their workforces had extended the IRT's lines up the east side of Manhattan to the Bronx and down the west side to the Battery. In a historic first that made New York one city in fact as well as in formalities, new tunnels under the East River connected the IRT to the BMT's new lines in Brooklyn and Queens. With the completion of the new subway lines in 1918, four of the five boroughs were connected to the city's rapid transit network.

Despite the new system's remarkable success, the trains were still crowded, many communities were still not served, and demands continued for the construction of still more subway lines. But the old disagreements continued about who would own them, who would build them, where they would be built, and whether they should be connected to the existing privately owned system. In a typical quarrel between the city and the rest of the state, there were even disputes about who should make these decisions—the solidly Democratic politicians in City Hall or the mostly Republican politicians from upstate who dominated the state government in Albany. Eventually, the arguments were resolved with the founding of a third entity, the Independent Subway Line (IND). In 1925, the ground was broken on new subway routes, including a line from Times Square in Manhattan to the neighborhood of Flushing in Queens, a line from Flushing to Main Street in Queens, and a line from 14th Street to Williamsburg in Brooklyn.

The subways did more than take passengers from their homes to their jobs. In very tangible ways, the subways created, united, and democratized New York City. Knitting together every borough except Staten Island (which could be accessed only by ferries from Brooklyn and Manhattan), the subways made it possible for people from Manhattan,

Brooklyn, Queens, and the Bronx to consider themselves citizens of the same city. Entirely new neighborhoods—for instance, Flushing in Queens and Kingsbridge in the Bronx—sprang up near new subway stations. Thus the subways precipitated some of the greatest real estate booms in history and literally laid the groundwork for New York City's long-term growth and prosperity.

Even more importantly, the subway system has always been a profoundly democratic institution. While the wealthiest New Yorkers may have traveled by taxis and limousines, people from almost every background and walk of life rub shoulders in subway cars. Thanks to the subways, it is possible for young people from working-class families from the farthest reaches of Brooklyn and the Bronx to commute to City College in Harlem—and to begin their journeys into the middle class. Even the poorest people could afford to take the trains to municipal museums, public libraries, public concerts, sporting events, patriotic parades, protest marches, and political rallies. For decades, all it cost to travel from Kingsbridge in the Bronx to Coney Island in Brooklyn was one nickel each way — the famous "five-cent fare." For the city and its citizens, the subways were one of the best bargains in history.

"The construction industry is a different world from where we were in the past. The business was very different then. We took huge risks. We lived from hand to mouth. There was a lot less planning and almost no safety, like the wild west."

Robert Koch, *President/CEO, Skanska Koch*

Ashokan Reservoir (right)
August 5, 1914.
© New York City Department of Environmental Protection.

Manhattan Bridge (opposite page)
Marine terminal in the foreground, bridge under construction in the background. March 23, 1909. Courtesy United States Library of Congress Prints and Photographs Division.

Water for the Metropolis

Providing a steady supply of drinkable water for a growing metropolis, a burgeoning population, and bustling businesses has always been a challenge for New York City. Manhattan and much of Brooklyn, Queens, and the Bronx are surrounded by the East River, the Hudson River, and the Harlem River, waterways filled with undrinkable brackish water—water that is saltier than freshwater but not as salty as seawater. As the city's residential, industrial, and commercial areas continued to expand during the nineteenth century, many sources of freshwater became polluted, much of the water supply became unsanitary, and epidemics of cholera and yellow fever afflicted New Yorkers. Without an adequate water supply, the city was also vulnerable to fires, including the "great fire" of 1835, which burned down much of Manhattan.

In order to address these needs and avert problems, the city's water supply system grew from several wells on Manhattan Island to a system of aqueducts, tunnels, and reservoirs stretching from far upstate to the five boroughs. Building, maintaining, repairing, and expanding this network has required extraordinary feats of engineering, construction, leadership, and teamwork.

During the nineteenth century, New York City built two aqueducts to bring water to Manhattan from northern Westchester County. In a project that was completed in 1842, the Croton River was dammed, and an aqueduct was constructed that could carry 90 million gallons of water a day from the Croton Dam through Westchester County and the Bronx, over the Harlem River, and into reservoirs in Manhattan. This water supply system came to be called the Old Croton Aqueduct. Following a similar route, the New Croton Aqueduct, with three times the capacity, was opened in 1890 and fed into the same reservoir.

One reservoir, where Bryant Park now is, was pulled down a few years later; a second reservoir was drained to become Central Park's Great Lawn in 1925; the third, now called the Jacqueline Kennedy Onassis Reservoir, still stands but was decommissioned in 1993.

As the twentieth century began, the newly consolidated metropolis decided to develop the upstate Catskill region as an additional source of water supply. Newly created by the state legislature in 1905, the Board of Water Supply made plans to transport water from the watersheds in the Catskill Mountains to New York City. The Catskill Aqueduct begins at the Ashokan Reservoir, goes south, sometimes underneath the Hudson River, and continues through Westchester County to the Kensico Reservoir then to the Hill View Reservoir in Yonkers from which City Water Tunnels 1 and 2 bring water to New York City. Beginning in 1907 under Mayor George B. McClellan, Jr. (whose monument stands atop Winchell Hill, near the site of the big dam), the 92-mile system was completed in 1927 at a cost of $162 million and 283 lives. It is capable of supplying 650 million gallons of water a day.[8] To this day, New York City could not survive without it.

Many GCA members worked on these projects, including Winston & Company, MacArthur Brothers Company, Blaw-Knox Company, Mason & Hanger Company, Locher, Dravo Corporation, Pittsburgh Contracting Company, Grant Smith Company, T. B. Bryson, American Pipe & Construction Company, Bradley Contracting Company, Bradley-Gaffney-Speers, Holbrook, Cabot & Rollins Corp., Rice & Ganey, Inc., J. F. Cogan Company, H. S. Kerbaugh, Inc., Ulen & Company, Degnon Contracting Company, Snare & Triest Company, Thomas Crimmins Contracting Company, T. A. Gillespie Company, and Thomas McNally.

The building of the aqueduct required huge and heroic efforts. Imagine thousands of workers in sweat-streaked shirts and floppy felt hats. Huge blocks of cut stone are dangling from wooden derricks. Steam-snorting locomotives are dragging rubble-heaped gondola cars along narrow ledges. Excavating machines on tracks are tearing trenches with giant clam-shaped jaws. In the distance, over the clang of hammers and spades and the whinnies and brays of horses and mules come the muffled reports—and acrid after-scents—of underground explosions. That is what it must have been like to watch the Catskill system's giant Kensico Dam go up.

All in all, as Professor Charles Merguerian of the geology department at Hofstra University has written, the Catskill system involved 67 shafts varying in depth from 174 to 1,187 feet; three masonry dams; several miles of earth dikes; and more than 163 miles of aqueduct, including 28.5 miles of grade tunnel, 35 miles of pressure tunnel, 6 miles of steel pipe siphons, and 39 miles of pipe conduit.[10] Not since the pyramids—or the mighty Roman aqueducts—had so much raw energy been expended to move so much dirt and stone.

Completed in 1917, City Water Tunnel 1 begins at the High View Reservoir in Yonkers and extends through the Bronx and crosses the Harlem River to Manhattan and the East River to Brooklyn. Covered in most places by at least 150 feet of solid rock through its course in the city, the tunnel is from 200 to 300 feet deep. Together with a second city water tunnel that was built two decades later, this system still serves the needs of a thirsty city.

Bridges and Tunnels

After the first Model T Ford rolled off the first assembly line in 1908, New York's residents and businesses started switching from horse-drawn vehicles to automobiles and trucks. In 1917, in an event that marked a tipping-point between the horse and the car, the last horse-drawn streetcar ran from lower Manhattan to Central Park South, and there were as many cars and trucks as horses on the city's streets. To accommodate this traffic, new streets, highways, bridges, and tunnels were built.

Completed in 1927, the Holland Tunnel was the first roadway tunnel crossing between New York City and New Jersey. Jointly funded by New York State and New Jersey, the tunnel went from Canal Street in lower Manhattan under the Hudson River to Jersey City. To oversee this task, the two state legislatures established two independent commissions that enlisted some of the most eminent engineers of the era, including General George W. Goethals, once chief engineer of the Panama Canal, Colonel William J. Wilgus, who had designed Grand Central Station, and Clifford M. Holland, who had designed East River tunnels for the IRT subway.

This would be the longest underwater tunnel in the world. Given the technology available at the time, there were many unknown factors. The engineers eventually agreed to include the traditional steel rings, as well as concrete, to provide the necessary strength and durability for the structure. In order to ventilate the tunnel, 3.6 million cubic feet of air were pumped every minute into a duct under the roadway.[11]

Holland Tunnel (right)
Traffic at New Jersey Entrance Ramp.
June 24, 1935.
Courtesy Port Authority of New
York and New Jersey.

Construction of Holland Tunnel (below)
Sandhogs in South Tunnel stand in front
of drum of east wall, circa 1927.
Courtesy Municipal Archive Collection.

From there, the air would be released through narrow slots above curb level. Meanwhile, the exhaust would be pulled out of the air via a duct that ran between the roadway ceiling and the tunnel tube's upper rim. The result: the air in the tunnel, as *The New York Times* reported a year after it opened, was "sweeter than many city streets."[12] GCA members the Carleton Company, William J. Fitzgerald, John Meehan and Son, and Booth and Finn all helped to build the Holland Tunnel.[13]

On November 12, 1927, President Calvin Coolidge pressed a gold and platinum telegraph key in the White House to open the tunnel. From the beginning, the volume of traffic exceeded all expectations, and the tolls that were collected quickly surpassed the tunnel's cost of construction. Soon truckers and other motorists demanded more tunnels. By the end of 1928, New York and New Jersey's Bridge and Tunnel Commissions each formally requested that the states approve a new tunnel.

In another ambitious endeavor, work began in 1927 on a suspension bridge across the Hudson River, connecting the Washington Heights neighborhood in upper Manhattan to Fort Lee in New Jersey. Because the bridge brings together two places that General George Washington used as fortified positions during the Revolutionary War—Fort Washington in New York and Fort Lee in New Jersey—it was named the George Washington Bridge. With Cass Gilbert as the architect and Othmar Ammann as the chief engineer, the George Washington is the fourth largest suspension bridge in the nation, with two levels and a total of 14 lanes. It was constructed by GCA members Foley Brothers, Inc., Arthur McMullen, George M. Brewster and Sons, Inc., Corbetta Concrete Corp., Cornell Contracting Corp., Rusciano and Son Corp., and Senior & Palmer, Inc.[14]

Meanwhile, the new Port Authority of New York and New Jersey built two bridges from Staten Island to New Jersey, both of which opened on June 29, 1928. The Goethals Bridge ran from Howland Hook on Staten Island to Elizabeth, New Jersey, and the Outerbridge Crossing ran from Staten Island's southwestern tip to Perth Amboy, New Jersey. GCA members Slattery Contracting Company and Frederick Snare Corporation were involved with building the Goethals Bridge and the Outerbridge Crossing.[15]

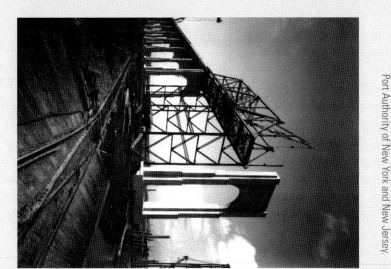

Outerbridge Crossing (below)
Laying reinforcing steel on the Outerbridge Crossing, Port Authority Approach. Courtesy Port Authority of New York and New Jersey.

Railway Links to the Rest of the Nation

The same transforming technologies that allowed New York City to build its subway system and the vehicular tunnels under the Hudson also allowed passenger trains to run between New York and New Jersey. As long as trains were powered only by coal-burning engines, it was impossible for New York to build subways or for railroad trains to travel by tunnel from the mainland to Manhattan Island. The smoke from the engine cars would have filled the tunnels, making it impossible to breathe safely or see ahead while traveling underground or underwater. But, with the invention of the electric locomotive and the electrification of sections of main-line track, it was possible for subways and passenger railways to travel through tunnels.

During the first two decades of the twentieth century, two of the nation's leading railroads, the Pennsylvania Railroad and the New York Central Railroad, built lines connecting Manhattan to New England, Long Island, New Jersey, and the Middle Atlantic and Southern states. The Pennsylvania Railroad drove a tunnel under the Hudson River, providing for a direct trunk-line passenger connection to the North American mainland. The New York Central Railroad tunneled under Park Avenue to the new Grand Central Station. Completing its system, the Pennsylvania Railroad developed a rail line through Brooklyn and Queens and across the Harlem River, over Wards and Randall's Islands, to the Bronx and New England, on the most expensive track ever before built and over the heaviest and most expensive bridge, the Hell Gate Bridge, ever built up to that time. GCA members Arthur McMullen, Degnon Contracting Company, Gaffney Gahagan Construction Corp., H. S. Kerbaugh, Henry Steers, Inc., and O'Rourke Engineering Construction Company helped to build the tunnels to Penn Station.[16]

These two railroads built two great railroad station buildings in midtown Manhattan, only 12 city blocks away from each other. Serving as the entrances to New York City for millions of travelers, the two stations were designed to resemble classical Roman and Beaux Arts public buildings.

Opened in 1910 at West 33rd Street and Seventh Avenue, Penn Station was modeled after the ancient Roman baths of Caracalla. Opened in 1913 at East 42nd Street and Park Avenue, Grand Central Station included an ornate 40-foot statue of mercury, the Roman god of speed. As two of the city's most magnificent structures, these stations reinforced New York's reputation as a world-class metropolis in the tradition of Rome, London, and Paris.

Meanwhile, a new commuter line began taking workers from New York City across the Hudson to New Jersey. In 1908 and 1909, the Hudson and Manhattan light-transit lines began running the two "PATH tubes" from lower Manhattan to New Jersey.

The Port of New York

Once New York City was consolidated, the new metropolis boasted the busiest harbor in the world with 578 miles of shoreline. Forty-seven percent of the nation's shipping passed through the narrow strait between Staten Island and Brooklyn to be loaded or unloaded at the city's docks or floated across the harbor to or from the railheads in Staten Island and New Jersey. New opportunities beckoned with the opening of the Panama Canal and the New York State Barge Canal System.

But New York City's role as the nation's leading port was challenged by several developments. Competing ports were growing up and down the eastern seaboard. Technological developments were emerging, and New York's long and narrow finger piers were becoming obsolete. Longer ships were being built so when they were tied to the old docks of New York, their sterns hung dangerously into the channel. Mechanized derricks and waterproof warehouses could not be accommodated on narrow piers. And the national rail network was cut off from the docks by a wide river on one side and, on the other side, by a narrow street and a welter of privately held, underdeveloped properties, near congested streets. Therefore, the railway cars couldn't reach the ships.

Built by GCA members George C. Rogers and Spearin Preston & Burrows,[17] the Chelsea Piers comprised a vast new steel and stone complex of covered docks, which was developed as one of the first efforts to provide more terminal capacity, particularly for passenger traffic. Designed by Warren & Wetmore (who also provided the architectural design for Grand Central Station), the project was begun in 1901 and completed in 1910. Still, most of the port's problems remained unaddressed.

Meanwhile, on some shore front lots in south Brooklyn, a young entrepreneur named Irving Bush was devising new ways for a port to do business. With the construction work beginning in 1902, the Bush Terminal complex was the first facility anywhere in the world to provide a seamless interface between ships, rails, and trucks and between transportation facilities, warehouses, and factories. With the most modern, mechanized, temperature-controlled, and fireproofed conditions, businesses could have almost all their needs filled, from insurance to marketing to meals—all in one location.

By 1909, the Bush Terminal was already so successful that the city government was hatching plans to take it over. By 1918, when the United States had entered World War I, the Bush Terminal was so essential to the entire Allied war effort that the federal government did take it over for the duration of the conflict. Alongside the terminal, the federal government constructed a reinforced concrete clone, the Brooklyn Army Terminal, which GCA member Post & McCord, Inc. helped build.[18]

Meanwhile, some efforts were underway to develop Jamaica Bay in Queens as New York's port of the future. Municipal officials convinced the federal government to spend some $6 million dredging new shipping channels to reach the modest new piers it built in Canarsie and Mill Basin in Brooklyn.[19] But ultimately these efforts came to naught. By mid-century, the Jamaica Bay shipping lanes were as filled with sand and silt as they had ever been and the shoreline which was never used had been turned into parks.

Yet another effort to bring the port into the twentieth century resulted from World War I. Mayor John F. Hylan argued that the wartime transportation crisis made new piers necessary on Staten Island, which, with the Kill Van Kull rail bridge to New Jersey, offered New York City its only direct rail-freight connection to the rest of the continent to the west and south. The city proceeded to build new piers in Stapleton on Staten Island's north shore. Completed after the war was won, the piers soon became hopelessly outmoded—too short, too narrow, and inadequately connected to onshore support systems. Therefore, they were never used beyond a fraction of their capacity. GCA members Terry & Tench helped to build the Staten Island piers.[20]

Creation of the Port Authority, 1921
Courtesy Port Authority of New York and New Jersey.

Con Ed East River Plant (right)
Waterside Plant
addition under construction.
Courtesy Con Edison of New York.
March 1, 1926

Con Ed East River Plant (below)
East River Station,
Courtesy Con Edison of New York. October 15, 1926

**Queens Boulevard-Gosman Street,
Flushing Line (opposite page)**
Flushing line under construction.
Courtesy New York Transit Museum.
November 24, 1914.

Electricity, Natural Gas, and Telephone Service

Although most of the city's stoves, boilers, and engines were still fired by coal, New York City's electrification was completed during the first decades of the twentieth century. The numbers of electric rate payers rose steadily from 1909 on, as most of the city's businesses and households plugged into the power grid. The city's major electric utility—the New York Edison Company (later called Consolidated Edison, or, simply, Con Ed)—scrambled to build enough new generating capacity to keep up with the demand, and to string enough new cable below the streets to carry it. The Bell Telephone Company also began to approach the point where its New York City demand was saturated, as its wires also wended their way below ever more streets.

The increasing use of electricity and natural gas required new construction projects. By 1909, the New York Edison Company had generating plants in all five boroughs. One of the largest was the Waterside Plant which extends from 34th to 41st Streets in Manhattan, which opened in 1901 and stayed in operation until 2005. This plant was the utility's largest until 1925, when the Hudson River plant on the southern edge of the Brooklyn Navy Yard opened and construction began on the Hell Gate station. Another large generating facility, on the East River at 14th Street in Manhattan, opened in 1926.

During this period, the gas used for cooking was still rendered from coal at several dozen gas plants. The major gas producing facilities constructed during this period included the huge Astoria plant (on which construction began in 1903 and was completed by 1913), the Hunts Point plant (1926), and Brooklyn Union Gas's Greenpoint/Newton Creek plant (completed in 1928). The gas from these facilities was stored in huge round silos whose flexible tanks rose and fell like lungs inside their metal outside skeletons as the gas volume inside them expanded or contracted. Local landmarks in neighborhoods across the city, the last of these brightly painted vessels was demolished in 2001.

World War I

World War I came as a shock to the city's economy, its social fabric, and its physical infrastructure. While construction came to an abrupt halt, virtually all the men and material shipped abroad to support the war effort flowed through the Port of New York. The unbroken line of railcars from the interior of the continent gridlocked the freight yards on the New Jersey shore. The steady stream of car floats and lighters hauled across the harbor nearly choked New York's piers.

The problem was national in scope, but it ended up in New York City, the nation's gateway to the rest of the world, further straining an already inadequate set of physical facilities. In the winter of 1918, when a hard freeze locked the harbor in ice, the city experienced its first acute "coal famine."[21] With storage space within the city limited to only a few days supply of coal, boilers went cold in homes and businesses all over the city. Police stations served as emergency shelters for tens of thousands of people who had no other way to keep warm. To conserve fuel, businesses and schools were ordered to close on specified days. Because they were seen as less than essential, taverns had to close. It was said that some people got married just to stay warm.[22]

New Yorkers learned how frail the city's physical infrastructure still was. There were new demands for planning—and for institutions capable of meeting the metropolitan area's requirements for infrastructure. Soon after the war was won and the crisis subsided, two new organizations, with distinct but complementary goals, were founded and went to work.

The Port of New York Authority and the Regional Plan of New York and Its Environs

Founded in 1922, the Regional Plan Association could call upon others to take action. Established in 1921, the Port Authority could take action, including selling bonds and using the money to build projects. Together, they helped to chart a course for the region's physical future.

The Regional Plan Association's first plan, released in 1929 just before the great stock market crash, outlined an interconnected web of motor highways for passenger cars and rail lines for freight. The Port Authority eventually built the highways, together with airports, port facilities, and skyscrapers.

Remarkably, New York would build these facilities in the midst of the Great Depression, World War II, and the aftermath of these cataclysmic events. Once again, New York would demonstrate its extraordinary resilience. All those who envisioned, designed, and built the great city's economic, social, and physical structures early in the twentieth century had done their work well.

Horse-Drawn Fire Engine (left)
Horse-drawn fire engine at 72nd Street and Broadway, New York City, racing to a fire. Photograph, ca. 1910. © Bettmann/CORBIS

Hutchinson River Parkway Extension Construction (right)
October 16, 1940. Photographer Rodney McCay Morgan. Courtesy MTA Bridges and Tunnels Special Archives.

Map of First Regional Plan, 1929 (below)
Courtesy Regional Plan Association.

THE REGIONAL HIGHWAY SYSTEM

KEY PLAN
FOR
REGIONAL HIGHWAY ROUTES

SCALE IN MILES

LEGEND
Metropolitan Loop
Other Routes
Metropolitan By-Pass

FIG. 12

MAY 1928

Digging: Going Down

"What goes up must come down."

When it comes to building skyscrapers, bridges, or elevated highways, that old saying could be modified: "What goes up must first go down." Before putting up the steel frames for office towers or spinning the cables for suspension bridges, builders need to dig the holes for the foundations that anchor these structures.

Most New Yorkers and visitors to the metropolis see only half the marvels of modern construction. Every day, we see the noble granite piers of the Brooklyn Bridge, the majestic steel skeleton of the George Washington Bridge, the Empire State and Chrysler buildings that almost block the sun, and the power plant stacks whose plumes blend with the clouds. But we don't see what lies beneath the sidewalks or appreciate the effort and ingenuity it takes to secure the skyscrapers and other high-rise construction projects.

What's underneath the city streets? Of course, there are the systems that New Yorkers have created: the subways, sewers, water mains, gas pipes, and electric lines. Beneath what people have built is what nature has bequeathed us: New York's bedrock. The city is blessed with unparalleled layers of crystalline rock—schist, gneiss, and dolomitic marble of varying thickness that sink hundreds of feet below the five boroughs.

George Washington Bridge (above)
George Washington Bridge, cutting the access route from the west, 1928. Courtesy Trentonian Library.

"In the past, construction was much slower than today, not as well mechanized. Many of the pictures of work that took place in the 1920's and 30's show horse drawn trucks. The work was more difficult and dangerous than it is today. When you look at the way the workers were dressed, they did not have safety equipment. The safety structures for protecting workers from falling did not really exist."

John Donohoe, *Chair, Moretrench Corp.*

Brooklyn Bridge (right)
View of the Brooklyn Bridge and the Manhattan skyline, New York, New York, 1930s. Photographer Frederic Lewis. Courtesy Hulton Archive, Getty Images.

George Washington Bridge (above)
On the New Jersey side, before a blast. Courtesy Trentonian Library.

Without such a firm base of bedrock, the skyscrapers, bridges, and other tall structures could not have been built. Nor could New York City have so high a concentration of population and commercial activity. Indeed, to trace the course of the layers of bedrock is to locate the areas on which skyscrapers and similar buildings have risen.

But bedrock isn't all that's below the streets, subways, and sewers. The city has a 578-mile shoreline, and the water has shaped the gravel, sand, and quicksand of clay and silt and landfill that lie on top of and alongside the bedrock. Together with the great opportu-

nities offered by New York's harbor and rivers, the waterways also present extraordinary challenges.

Over the past hundred years, the route to the city's bedrock has been dug, blasted, and bored in many ways. Different methods have been used, depending on what is located above and below the excavation and what techniques had been invented at the time. At the turn of the last century, the pick, spade, and wheelbarrow still played an important role in prying material from the ground and hauling it away. Since then, human muscle has never been completely replaced, but the needs have diminished.

63rd Street Tunnel (above)
Mucking operation in the shaft area in
Queensbridge Park, Queens, for the
construction of the 63rd Street Tunnel.
February 4, 1971.
Courtesy New York Transit Museum.

Even a hundred years ago, explosives played an important part in breaking rock into pieces small enough to be removed. Over the years, steam, pneumatic, and electric drills replaced hand-powered tools. Picks soon gave way to air-powered jackhammers. Steam and diesel-powered excavators — mounted on rails, tracks, and trucks — became hugely more efficient.

Combining mind and muscle, fantastic feats of excavation were required to build most of New York's landmarks. For instance, Rockefeller Center conjures up images of a mountain created by humans that extends from the foothills of Radio City to the summit of the Rainbow Room, with peaks and cliffs built from many thousands of tons of concrete and steel. But the buildings also required a remarkable foundation. The builders used dynamite to blast their way through 70 feet of bedrock. In the process, they removed more material, by weight, than was used in the buildings that rose over the holes they created. The bedrock was removed, one vertical "lift" at a time, by drilling down on horizontal "benches." Long channels cut by pneumatic drills were packed with dynamite, then covered by heavy metal mesh mats to contain the fragments that would be blasted loose by the explosion.

Digging Below the Waterline

Difficult as this excavation work was, blasting through solid, dry bedrock is relatively simple compared to digging below the waterline. The device that does that—a steel, wood, or concrete box, open at the bottom and filled with pressurized air to keep the water out—is called a caisson. As workers inside the caisson dug down through the material overlying the bedrock, the weight of the box—and of the concrete, steel, and stone piled on top of it—kept pushing the box farther into the ground. An airlock between the working chamber and the shaft entrance allowed men and material to enter and men and muck to emerge without causing a loss of pressure at the working face. It also allowed workers to adjust gradually to the changes in air pressure, so that they could avoid a case of the painful and potentially permanently debilitating or fatal "bends" caused by too sudden changes in pressure.

BRT Broadway Line (above)
In blasted tunnel excavations, the skips (this kind of skip was called a "battleship") were hoisted up a shaft by a steam drum mounted in the headhouse. BRT Broadway line, 60th Street tunnel. November, 1917.
Courtesy New York Transit Museum.

BRT Broadway Line (right)
BRT Broadway line around 60th Street and Central Park. October 6, 1915.
Courtesy New York Transit Museum.

Invented in England by Thomas Cochrane in 1830, pressurized caissons were first used in New York to sink the piers for the Brooklyn Bridge (which was built from 1869 through 1883) and to begin construction in 1874 of the cross-Hudson tunnels through which the PATH trains now run. After GCA member Daniel A. Moran invented an airlock that used a relatively narrow steel cylinder, caissons could be used to excavate foundations for inshore buildings. The first "pneumatic process" foundation was built in New York in 1892 for the Manhattan Life Insurance Company Building. Within 20 years, caissons were used to build 50 foundations in New York, as well as in other North American cities.

Despite early efforts to protect workers from the bends, many workers who helped build New York were crippled or killed by "caisson disease." The work was so arduous that physically fit men could work for only a few hours under air pressure in a horizontal tunnel and for less than an hour in a vertical caisson shaft. Fortunately, advances in boring and pile-driving technology now allow caissons to be sunk without requiring laborers to work inside them.

In addition to digging underground or beneath the waterline, there's another complex and arduous task: hauling away the vast quantities of soil and rock that are removed from the foundations and shafts beneath every structure. In 1909, wheelbarrows and mules still played an important role in moving "muck" (the term for any material blasted or dug from an excavation). Buckets suspended from derrick-mounted, steam-drum-powered pulleys were another early means for removing muck from a shaft. From there, the waste materials often went into a "muck car," driven by a steam-powered locomotive.

With the development of more powerful gasoline engines after World War I, it became possible to drive heavy trucks up relatively steep grades. This advance made it easier to remove the muck from excavation sites. GCA member Patrick McGovern is credited as the first contractor to use ramps to get trucks into and out of the sites. Thus, the muck could be loaded directly onto the same vehicle that would carry it all the way to its disposal destination. McGovern's new method played an important role in building the IND subway, which was begun in 1925.

Reusing the "Muck"

Sometimes, the muck could be reused on-site in the next phase of construction. This was done in building New York's aqueduct system. Chunks of excavated rock were used as pieces in the "cyclopean masonry" of the dams. Crushed excavated rock was used as aggregate in the concrete that was mixed on-site to line the water tunnels.

If the muck wasn't reused on-site in the next phase of construction, it had to be loaded for transport and taken someplace where it could be disposed of. Because this was so costly—and because of the muck's potential use as fill—builders and owners had strong economic incentives to dump their loads as near as possible to the excavation sites. Public agency construction sponsors had strong incentives to use the fill to make new publicly owned land.

For instance, a huge amount of soil and rock was removed from a nine-mile stretch of the Eighth Avenue IND subway in Manhattan in 1925. If it had been piled on a standard city block (about 200 by 600 feet), it would have formed a 900-foot-tall cube. Instead, the city had this material dumped into the Hudson River north of 72nd Street on the west side of Manhattan to create 94 acres of new shore front in Riverside Park.

Nearly half a century later, the 1.18 million cubic yards of material excavated from the "bathtub" that would form the foundation of the World Trade Center was managed with even greater care. After dredging the mud and silt from the river bottom, GCA members Spearin Preston & Burrows and Horn Construction built a "box" in the Hudson River into which GCA members Slattery, Gull, Poirier and McLane, Tully & DiNapoli, and Spencer, White and Prentis put the excavation spoil from the World Trade Center in 1968.

The material was piled inside a watertight cofferdam (a dry enclosure underwater) composed of 40 cylindrical sheet piles to form the 23.5 acre Battery Park City.

World Trade Center (above)

Construction of Cellular Sheet Pile cofferdam for the cellar dirt from the construction of the World Trade Center. The 23.5 acres of new land created by the landfill project will later become Battery Park City. Construction by a joint venture of GCA members Spearin, Preston & Burrows and Horn Construction Company. Courtesy Spearin, Preston & Burrows.

> "Water tunnel jobs were my favorite because of the complexity of the project and satisfaction gained upon completion of the work."
>
> Thomas King,
> Schiavone Construction Co., Inc.

Tunneling: Going Sideways

Central Park, Manhattan (above)
IRT Contract 1 line. Note the rolling platform or "traveler" in the newly excavated tunnel. This traveler, which operates on rai s placed along the edges of the rock wall, is used to build the concrete sidewalks. Courtesy New York Transit Museum.

New York City's infrastructure includes 24 miles of vehicular and railway tunnels, 137 miles of subway tunnels, and 400 miles of aqueduct and water tunnels. That's more than enough mileage to provide a subterranean passage to Cincinnati. And that's not counting the city's 6,300 miles of sewers.

As early as the middle of the nineteenth century, builders started digging New York's tunnel network. The Croton Aqueduct, which opened in 1842, included 41 miles of tunnel that carried water from upstate watersheds to the heart of the city. In the late nineteenth century, the city began building an engineered network of large-bore collecting mains to intercept the common sewers that had simply discharged every side street's waste into the nearest water body. Initial construction of the trans-Hudson tunnels that would one day serve the PATH trains began in 1874.

As the twentieth century began, the pace of tunnel construction quickened. The construction of the city's first subway began in 1900. The boring of the Pennsylvania Railroad's tunnels from New Jersey under the Hudson to Manhattan—and under Manhattan and the East River to Queens—began in 1903. All in all, the great majority of the city's tunnel mileage was built within the past hundred years.

Because of the complexity and variety of conditions underground, every known tunneling method has been used in some part of New York City. As these methods have evolved, New York's tunnel builders have taken full advantage of the technological advances.

View of Church Street Construction Looking North from White Street (above)
Construction of the IND 8th Avenue Subway Line. August 8, 1928. Courtesy New York Transit Museum.

Trunk Sewer in Queens (below)
Construction of trunk sewer in Queens by GCA member Nicolas DiMenna & Sons, 1938. Courtesy New York City Municipal Archives.

There are six basic methods of tunneling: cut and cover, boring, shields, submerged tubes, drill and blast, and tunnel-boring machines. In choosing a method, builders mostly consider how deep the tunnel will go and what types of materials the builders are digging through — from soft clay to hard rock. Many times, a combination of two or more methods is used to build a tunnel.

The Simplest Method: Cut and Cover

Usually, the cut-and-cover method is quickest, easiest, and least expensive. However, this method is suitable only for tunnels that run close to the surface. The condition of the surface land must make it possible for a large trench to be dug. Based on the success of cut-and-cover in Budapest, William Barclay Parsons, the designer of New York's first subway, chose that method for building the first stretch of the IRT (which was constructed from 1900 through 1904). From 1913 through 1932, much of the excavation for the subway system was cut and cover, including the "dual contract" system, which combined the IRT with the Brooklyn Rapid Transit (BRT; this was later called the BMT—Brooklyn-Manhattan Transit) and the Independent system (IND). The method was also used for significant parts of the Catskill Aqueduct from 1907 through 1928 and the Delaware Aqueduct from 1931 through 1967. Some of the city's interceptor sewer network was also built with cut and cover.

the first subways, contractors did not have to work around already existing underground tracks. And New Yorkers had not already been outraged by the disruption to street life that subway construction caused. There was less knowledge and concern about safety, and the Occupational Safety and Health Act and similar state and local laws were decades away.

Here's how cut and cover worked: First, a trench was cut to the desired depth. Inside the trench, a longitudinal box was built whose floor was usually a concrete pad and whose walls and ceiling were generally steel girders and beams encased in concrete. Then the street surface was restored on top of the box.

During the first wave of construction, contractors did little to avoid disrupting street activity while the work progressed. They left their trenches open to the sky until the construction work was almost finished. And they did little to provide passage for vehicles or pedestrians.

By the time the second wave of construction began in 1913, public pressure made it necessary to provide a temporary deck as soon as the top level of the street surface had been removed far enough to provide enough headroom to work below it. Usually the deck consisted of wooden beams or steel sheets laid across the opening and supported by steel girders.

When contracts were issued for the IRT and IRT/BRT systems, much excavation was still done by hand. The contractors also ran steam shovels along street surfaces to reach into the cut below. Workers also blasted away any rock or other subsurface obstructions with DuPont dynamite. Why were builders allowed to use these crude and noisy methods which would be unthinkable today? Because these were

Second Avenue Subway (left)
Cut-and-cover method of construction for the Second Avenue subway from East 110th Street to East 120th Street in Manhattan. The work was performed by GCA members Cayuga Construction Corp. and Thomas Crimmins Construction Company. Aerial view of intersection of Second Avenue and 115th Street showing construction crane over opening in street and temporary planks covering the roadway. October 28, 1975. Courtesy New York Transit Museum.

Catskill Aqueduct (above)
After the war, laborers were in short supply. This flyer was used to recruit workers for the Catskill Aqueduct system in 1921.

Contractors were allowed to dig trenches on only one side of the street at a time so that traffic could continue to use the other lanes.

With the third wave of subway construction in 1925, excavation methods were considerably advanced. In the years after World War I, the high cost and scarcity of labor led contractors to maximize the use of mechanical devices. This was possible because of improvements in machinery, more powerful trucks, and a wider variety of excavators (which, by that time, were powered by diesel, gasoline, or electricity, as well as steam). Also, the increasing ingenuity in the use of ramps allowed workers to move excavation equipment in and out of pits and trenches.

Some recently retired contractors and workers remember how subways used to be built. For instance, Thomas King, a retired executive of the Schiavone Construction and a past president of the GCA, remembers: "In the subway work, some of our first jobs were what they call cut and cover. Out in Queens, we would dig down and put wood planking and a temporary structure over the roadway. People would ride on top and we would build a subway underneath."

Boring through Bedrock

Cut and cover wouldn't work for underwater tunnels. Nor would it work for the deepest tunnels or parts of the subway system that went too far below ground (for instance, beneath the heights of northern Manhattan). And the method was not practical where many layers of subways and other underground rail lines crossed (as in congested midtown and lower Manhattan).

In these situations, tunnel builders had to bore through whatever lay in their way. Depending on the time and place, different techniques were required.

Tunneling with a "Shield"

For tunnels that did not go through rock but were too deep for cut and cover, boring was done with a shield. Invented by the British engineer Marc Brunel in the course of tunneling through the silt, sand, and mud underlying the Thames in 1825, the technique changed relatively little during the next century. Usually, the shield was a metal cylinder slightly larger in diameter than the intended size of the finished tunnel; it resembled a tin can with both ends cut off.

Lincoln Tunnel (above)
Construction of concrete cradle for shield. December 28, 1934. Courtesy Port Authority of New York and New Jersey.

Lincoln Tunnel (below, left)
Rear view of shield entering rock tunnel in Weehawken showing temporary rirgs to take "push" of jacks. October 2, 1934. Courtesy Port Authority of New York and New Jersey.

When an unfinished tunnel was being dug through soft earth, the shield prevented the walls and ceiling from collapsing on the workers inside it. Jacks at the back of the shield pushed its front face forward so that its cutting edge kept digging into the earth wall ahead. A hood at the top of the shield projected a few feet beyond the rest of the front face, providing a protective overhang that kept rocks and soil from falling on the diggers below. While some workers dug with picks and shovels, other workers followed behind them, carrying the broken rock and soil out of the shield.

Each time the shield was pushed forward by the jacks at its back, another team of laborers began bracing the newly uncovered section of tunnel wall with curved plates of iron or steel. Bolted together, these plates formed a complete ring around the inside of the tunnel. When this new ring was bolted to the ring immediately behind it, the shield jacks were braced against it for another shove forward. With each shove pushing another ring's width ahead (generally about 2½ feet), a shield might advance about 8 feet during a three-shift day.

If a shield was used for a tunnel under water or in a porous soil beneath the water table, pressurized air, as in a caisson, was needed to keep the tunnel from flooding. Despite the protection offered by pressurization, there was always the risk of flooding. For instance, when crews from the George H. Flinn Company (which was a member of the GCA) were boring the Brooklyn end of the Brooklyn Battery Tunnel, there was sudden loss of air pressure. As the head wall began to collapse, the workers cool-headedly pushed empty cement sacks into the cavities, keeping the tunnel intact and saving their own lives. Unfortunately, such incidents occurred all too frequently any time that pressurized tunneling through unexplored strata was required. And tunnelers were not always so lucky in avoiding loss of life.

Tunneling with Submerged Tubes

The fourth basic type of tunneling is done with submerged tubes. The brilliant William Wilgus, chief engineer of the New York Central Railroad, was the first to use submerged tubes on a major scale with the Michigan Central Railroad's Detroit River tunnel in 1906.

Later, the tubes were used in the first subway line across the Harlem River between Manhattan and the Bronx and in the 63rd Street subway line between Manhattan and Queens.

This technique involves laying prefabricated sections of steel or steel-and-concrete tunnel tubes in a stabilized channel under a river or harbor. First, a trough is dredged in which the tunnel can be laid. Then, if necessary, piles are sunk inside the trough to stabilize and support the tube. Then the prefabricated tube sections are floated into place over the trough and lowered in by allowing the tubes to fill with water. Divers guide the tubes into place and insert pins that lock the tunnel sections together. When the tubes are aligned and securely joined, they are buried under a layer of cement that is pumped underwater through "tremie" (smooth steel) pipes. Finally, the water inside the tubes is pumped out, and all the interior joints are caulked. When it is feasible, the technique offers significant time and cost savings over traditional boring.

Whether a tunnel carried cars and trucks, trains, water, or sewage, unless the tunnel consists of prefabricated tubes, its interior was finished with concrete to provide long-term stability and waterproofing. If the tunnel was built using a shield (with steel rings installed as part of the shield-driving process), a layer of concrete was applied on the innermost steel ring. If the tunnel was blasted, curved steel plates—or steel arches or girders and beams—were installed to stabilize it permanently. Then concrete was applied with the use of portable forms (which were already almost exclusively made of steel by the 1910s).

Cast Iron Tubes (above, left)
Construction of wood lining in cast iron tubes into which concrete will be poured, 152nd Street and the Harlem River, July 11, 1913. Courtesy New York Transit Museum.

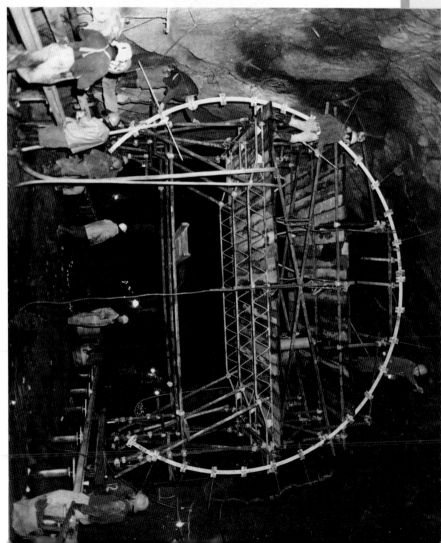

New York City Water Tunnel No. 3 (above, right)
Steel supports being bolted into subway tunnel, March 23, 1983. © New York City Department of Environmental Protection.

With contemporary tunneling done with a tunnel-boring machine, there no longer are steel arches or girders and beams for shoring. Instead, bolts are used to tie together rock layers that need stabilization, and concrete usually is sprayed (it is then called "shotcrete") onto a layer of mesh to provide long-term structural support and waterproofing.

Drill and Blast

Before tunnel boring machines were invented, the only economical way of excavating long tunnels through hard rock was drilling and blasting. Workers couldn't dig through hard rock, so they had to blast it away.

At first, gunpowder was used for drilling and blasting. Then safer and stronger explosives, such as dynamite, were developed. The drilling equipment includes pneumatically (compressed air) powered jackhammers and hydraulically powered jumbo drills. The "tunnel muck" (broken rock) was removed from the tunnel by rubber tire loaders.

First, the ground on top of the tunnel is reinforced by underground supports. Then workers drill holes in the ground to the tunnel's designated depth. Then they load the holes with dynamite or another blasting agent. When the explosive is detonated, the rock collapses, opening up the space for the tunnel. Then, the rubble is removed, new tunnel surfaces are reinforced, and a new area is detonated and cleared. Eventually, the bedrock has been blasted away, and a tunnel has been built.

Early in the twentieth century, drilling and blasting could be dangerous, but it was also straightforward. Parsons' contractors used this technique to tunnel under the deep crystalline rock of Washington Heights in upper Manhattan and through the mid-Manhattan bedrock around Park Avenue and 42nd Street. This method led to some fatal accidents (including the death of one of Parsons's key lieutenants), as well as some significant property damage (for instance, on Park Avenue, where there were many fissures in the rock). But, for the most part, the blasted-through rock provided a relatively stable opening through which a finished tunnel and tracks could be built.

Until 1913, blasting continued to be the most frequently used technique to tunnel in dry bedrock. By far, the greatest lengths of blasted tunnel are in the 92-mile Catskill and 84-mile Delaware Aqueducts. City Water Tunnels 1, 2, and a portion of 3 were also built using this method.

Tunnel-Boring Machines

In recent decades, tunnel-boring machines have replaced earlier equipment. These marvelous machines function a bit like Brunel's shield. But the digging and cutting at the head wall is all done automatically by rotating blades that tear through either rock or soil. The mucking out is done automatically by conveyor belt. The compressed air is replaced by a pressure-balanced cutting surface that exerts just enough force against the head wall to equalize pressure inside and outside the machine. And the workers sit at computerized consoles monitoring the progress of their machine through whatever lies ahead.

That's a far cry from the generations of workers who risked their lives digging, blasting, and building tunnels. And it's a sign of how far the construction industry has come in the first hundred years of the GCA.

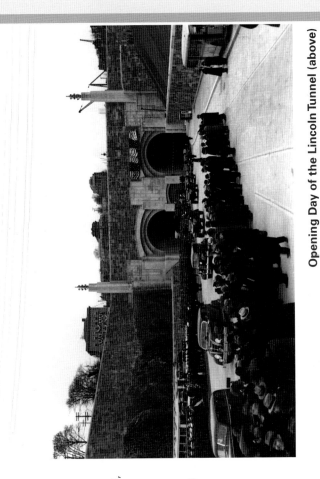

Opening Day of the Lincoln Tunnel (above)
Courtesy Port Authority of New York and New Jersey.

Timeline: 1930–1945

1930	1931	1932	1933	1934	1935	1936	1937
New York Central Railroad	Floyd Bennett Field	Metropolitan Avenue Bridge	Harway Avenue Bridge	Jackie Robinson Parkway (Interboro Parkway)	Triborough Bridge opens	Marine Parkway Bridge opens	
	George Washington Bridge	New Deal	Robert Moses appointed first citywide parks commissioner	Coney Island Sewage Treatment Plant opens	Henry Hudson Bridge	Mosholu Parkway	
	Bayonne Bridge	238th Street Bridge	Public Works Administration	West 215th St Incinerator	Grand Central Parkway	Pelham Parkway	
	Hook Creek Bridge	Rockefeller Center foundations blasted	Works Progress Administration	Zerega Avenue Incinerator	Henry Hudson Parkway, Riverside Park (New York Central tunnel)	Jacob Riis Park	
	Little Neck Bridge	Miller Highway (West Side Highway/Joe DiMaggio Highway)	New York Central West Side Track Elimination Project	East 60th Street Marine Transfer Station	East River Drive (Franklin D. Roosevelt Drive; FDR Drive)	Intercepting sewer from the Bronx to Wards Island	
	Eastern Boulevard Bridge	Reconstruction Finance Corporation	Mayor Fiorello H. LaGuardia	West 54th Street MTS	Interstate Sanitation Commission	Wards Island Sewage Treatment Plant	
	Fresh Kills Bridge	Northern State Parkway	Highline opens	Hamilton Avenue Marine Transfer Station	Consolidated Edison Co. of New York is created by merger of Consolidated Gas Co. with its subsidiary Edison Electric Co.	West 56th Street Incinerator	
	Cropsey Avenue Bridge	Eighth Avenue/IND Subway opens			Tunnel Authority created to build Queens-Midtown Tunnel	Flushing Incinerator	
	Centre Street Loop Chambers St. to Battery	Mayor John P. O'Brien			Orchard Beach	East 91st Street Marine Transfer Station	
	14th Street–Eastern District Line to Eighth Ave.				Intercepting sewer from Bronx and Manhattan, tunnels under rivers, to Wards Island	Lincoln Tunnel, first tube, opens	
	Cranford expands into Ready-Mixed Concrete				Water Tunnel No. 2 completed		
	US Supreme Court allows NYC to use Delaware River water; work begins on Delaware Aqueduct						

1938	1939	1940	1941	1942	1943	1944	1945
Westchester Avenue Bridge	North Beach/LaGuardia Airport	Mill Basin Bridge	Hutchinson River Parkway Bridge	Hamilton Avenue Bridge	Jamaica Sewage Treatment Plant	26th Ward Sewage Treatment Plant	War ends in Europe and Japan
52nd Street Marine Transfer Station	Cross-Bay Bridge (Veterans' Memorial Bridge)	Midtown Highway Crossing Bridge	World War II	Second Avenue El		Delaware Aqueduct completed	Municipal Asphalt Plant
	Flushing Bridge (Northern Boulevard Bridge)	Ninth Avenue El	Fourth Avenue El, Fulton Street El, Fifth Avenue El	Moran Co.'s fleet of 112 towing vessels includes 49 V-4 tugs loaned by the federal government to assist in the war effort; more than half of all the men and material shipped abroad will go out of NY			Lincoln Tunnel, second (north) tube opens
	Kosciuszko Bridge	Belt Parkway	Hutchinson River Parkway (Bronx portion)				Con Ed closes last gas manufacturing plant
	Bronx-Whitestone Bridge opens	Cross Island Parkway	Gowanus Parkway opens	East River Drive opens from 53rd to 92nd Street			
	Whitestone Express-way Bridge	City unifies subway system by acquiring BMT and IRT		City Island–Hart Island Sewage Treatment Plant			
	World's Fair	Sixth Avenue Subway opens					
	Sixth Avenue El	West 59th Street Marine Transfer Station					
	Whitestone Express-way	Queens-Midtown Tunnel opens					
	Bowery Bay Sewage Treatment Plant in operation						
	Tallman's Island Sew-age Treatment Plant opens						

1930–1945:

In Difficult Times, Building the Foundation for the Modern Metropolis

Triborough Lift (above)

View of the Manhattan tower of the Triborough Bridge's lift span taken from the lift span's Randall's Island tower. Photographer unknown. March 26, 1936. Courtesy MTA Bridges and Tunnels Special Archives.

Triborough (opposite page)

Construction on Randall's Island. The bridge's Bronx span is at right and its lift span to Manhattan at upper left. Part of the New York Connecting Railroad Bridge is visible at the bottom. Photographer Fairchild Aerial Surveys, Inc. 1936. Courtesy MTA Bridges and Tunnels Special Archives.

Building New York City's infrastructure has always been an act of faith that tomorrow's growth will justify today's investments. Betting on the future never involved more courage, foresight, and imagination than during the Great Depression and World War II. In the midst of economic collapse and international conflict, New York City built much of the physical plant—the bridges, highways, tunnels, airports, and sewage treatment plants—that makes the modern metropolis possible. By 1945, the great construction projects of the previous decade and a half had built the foundation for the expanding middle class and the enduring postwar prosperity, in New York City and throughout the nation. It was an accomplishment worthy of the members of the "Greatest Generation," who worked their way out of the Great Depression, won World War II, and built a better future for their children and grandchildren.

The Great Depression began with the crash of the New York Stock Exchange, and New York led the way as the nation lifted itself out of the economic doldrums. During less than a week on Wall Street in October 1929 the stock market collapse plunged the nation into a deep and lasting decline. In the New York metropolitan area, as millions of dollars in paper wealth turned to ashes, businesses went bankrupt, banks collapsed, workers lost their jobs, and state and local tax revenues declined. Most residential and commercial construction ground to a halt, and building contractors and craft workers shared in the hard times.

While the construction industry declined, building and repairing the city's infrastructure was actually more essential than ever. During the high-flying but shortsighted 1920s, many of the city's parks, roadways, and other public facilities had fallen into disrepair. With a growing population, bustling businesses, and new means of transportation from the automobile to the airplane, new infrastructure was needed. Much of what a modern metropolis needs—for instance, sewage treatment plants—had not yet been built. Meanwhile, in the midst of economic collapse and mass unemployment, new construction projects—what public officials used to call "public works"—were also urgently needed to create jobs, encourage business activity, and send signals of hope at a time of despair.

"My grandfather started Tully Construction Co. in the 1920s and since the beginning we've played a large role in building the foundation of New York City. We were involved in building the original Triborough Bridge, the 1939 World's Fair, the George Washington Bridge, Grand Central Parkway, the foundation of the World Trade Center and clean up after 9/11. Wherever you go in NYC, you'll see a project Tully helped make happen."

Peter Tully, *President, Tully Construction Co., Inc.*

Queens Midtown Tunnel (opposite page)

View showing drilling carriage in position for North tunnel drift face. August 8, 1939. Photographer Voss Studio. Courtesy MTA Bridges and Tunnel Special Archives

Lincoln Tunnel (right)

New Jersey shield during assembly, showing the front or cutting edge. The steel cutting edge castings and the position of the working platforms may be clearly seen in this picture. Camera set up on cradle near the east net line of the ventilation shaft, looking west. GCA member Mason & Hanger Co., Inc. Courtesy Port Authority of New York and New Jersey.

But those with the resources required for large construction projects—banks and brokers, private investors, commercial developers, middle class consumers, and local tax revenues—all were depleted by the Depression. In order to plan, fund, begin, and complete the massive and innovative construction projects that New York needed, the city, state, and nation cried out for visionary leadership and unprecedented teamwork. Fortunately for New York City, New York State, and the nation, the leadership and the teamwork emerged.

Indeed, New York City provided the inspiring leadership—and the exemplary teamwork—that helped to revive America's economy. In the White House, President Franklin D. Roosevelt, who had previously served as Governor of New York, led the nation out of the Depression with programs like the Public Works Act and the Works Project Administration that funded major construction projects in order to generate jobs and boost business. In City Hall, Mayor Fiorello LaGuardia used his formidable talents as an inspirational leader, tough manager, and super-salesman to plan projects, gain federal funding, and make sure they were completed quickly and capably. At LaGuardia's right hand was Robert Moses, the legendary "master builder" and "power broker." Holding as many as 12 public offices at one time, Moses served for many years as New York City Parks Commissioner, head of the New York State Parks Council, and chairman of the Triborough Bridge and Tunnel Authority. During a career in public service that spanned four decades, Moses was responsible for New York City opening 658 playgrounds and the city, state, and public authorities building 13 bridges and 416 miles of highways. Among Moses' most important accomplishments were the Triborough Bridge, the Verrazano-Narrows Bridge, the West Side Highway, and the Long Island parkway system.[1]

All three leaders owed much to another former Governor of New York, Alfred E. Smith, who had hired and mentored Moses, recommended Roosevelt as his successor (although the two would part ways politically), championed public works projects during the 1920s, and became the public face for the major private project of the early 1930s, the world's tallest structure, the Empire State Building.

With funding from the Roosevelt administration, LaGuardia and Moses set about building an ambitious array of construction projects that helped to revive New York's economy while preparing the metropolitan area for the phenomenal growth that took place for three decades after 1945. These leaders asked all New Yorkers—business and labor together—to work together to build the future, and the General Contractors Association and the trades unions answered the call.

During this difficult but enormously creative and productive era, the construction industry developed new and improved ways of building public facilities. For instance, there was the switch from bagged to bulk concrete—an innovation whose apparent modesty was belied by the outsize effect it would have on making construction operations more efficient. Opening in 1931 and served by a fleet of ready-mix trucks, the Cranford Company's bulk-concrete batch plant on the Gowanus Canal in Brooklyn may have been the city's first such facility. Other new technologies that first made their appearance at the city's construction sites included the use of gravel packing in place of grout to prevent the subsidence of buildings during tunnel boring and the use of pre-cast concrete instead of cast-iron lining for subway construction. And New York City's contractors and workers began to use the now ubiquitous hard hat, which had first appeared during the construction of the Boulder Dam and the Golden Gate Bridge.[2]

Triborough Bridge (above)
View looking south from
Astoria, Queens. GCA member
Arthur McMullen Co. contractor.
November 14, 1931.
Courtesy MTA Bridges and
Tunnels Special Archive.

**Triborough Bridge
(above, right)**
Concrete pier construction for the
Triborough Bridge viaduct on Wards
Island is underway in the foreground.
In the distance, cable spinning on the
suspension span is taking place. May 8,
1935. Photographer unknown. Courtesy
MTA Bridges and Tunnels Special Archive.

**Queens Borough Bridge Plaza
(opposite page)**
Completion of the excavation for subway
construction under the Queens Borough
Bridge, Long Island City, NY. The men
here are shown gathered around the last
load of dirt to be removed. June 30, 1930.
Courtesy New York Transit Museum.

"The World of Tomorrow"

By hosting the 1939 World's Fair, New York City sought to restore faith in the future. While the fair's theme was "the world of tomorrow," the fair itself served as a spur for rebuilding the New York of the 1930s and 1940s. Shrewdly, Robert Moses used the process of preparing the fairgrounds, as well as the excitement surrounding the international exposition, to promote projects of all kinds, from highways to a new airport and sewage treatment plants. Many GCA members were involved in pre-paring the site for the World's Fair and constructing the buildings for the expositions, including Arthur A. Johnson, Necaro Co., Tully & DiNapoli, Andrew Catapano Co., Grow Tunneling, Slattery Contracting Co., and Chas. F. Vachris, Inc.[3]

In a tribute to the public investments of that era, what once was the site of the 1939 World's Fair (and the 1964 World's Fair as well) now is the Flushing Meadows Corona Park in the borough of Queens. But, in the early 1930s, the site was a garbage dump with almost three decades' accumulation of refuse and ashes. Starting with the cleanup of the future fairgrounds, New York moved ahead with other plans that would help the city host the international exposition, while also improving its infra-structure for years to come. In addition to ending the use of the Corona Meadows as an urban trash can, the city also stopped dumping garbage on Rikers Island, where the landfill reached a height of 140 feet, and stepped up its sewage treatment plan. Similarly, the need for speedy transportation to the World's Fair site encouraged the city to develop a network of highways and crossings for cars, taxis, buses, and trucks to drive from Manhattan to Queens.

Connecting Three Boroughs

"Make no small plans" was another watchword for that era. Together with the World's Fair, the biggest, boldest endeavor was the Triborough Bridge that connected Manhattan, Queens, and the Bronx. Of New York City's five boroughs, only the Bronx is on the North American mainland, while Manhattan and Staten Island are separate islands and Brooklyn and Queens are on Long Island. The Triborough Bridge was built at the points where Manhattan, Queens and the Bronx are closest together. It overlooks Hell Gate, the rough waters where the East River, Harlem River and Long Island Sound all come together, and two small islands, Ward's Island and Randall's Island. Work on the project had begun one week after the stock market crash of 1929, but, with the Depression depleting the city's resources, it halted in 1932 and resumed a year later after Moses had created the Triborough Bridge Authority and the federal government had provided a $37 million loan.

Construction of Henry Hudson Parkway (below)
Looking north from 79th Street across Riverside Park and construction of Henry Hudson Parkway transverse. Work performed by GCA member Nicolas DiMenna & Sons. Courtesy Municipal Archive Collection.

Hutchinson River Parkway Extension Construction (below, right)
Hutchinson River Parkway Extension construction.
Photographer: Rodney McCay Morgan. October 16, 1940. Courtesy MTA Bridges and Tunnels Special Archive.

Completed in 1936, the Triborough Bridge is an engineering and construction feat of heroic dimensions. Instead of being just one big bridge, it includes three long span bridges, several smaller bridges and viaducts, and 14 miles of approach highways, as well as parks and recreational areas. Five thousand building trades workers labored on the project site to construct the bridge, and, in an example of how public works promote economic growth, cement plants and steel mills reopened throughout the nation to produce the concrete, beams, and girders for the bridge. All in all, 31 million hours of work were required to build the bridge.[4]

At the time, it was an act of faith to assume that there would be enough traffic through depressed areas of upper Manhattan and the East Bronx and a mostly unsettled section of Queens to justify the huge project. But, in the years ahead, Queens became New York City's major middle-class bedroom community, as well as a center of business, and the Triborough Bridge knit the three boroughs together, just as the Brooklyn and Manhattan Bridges had made Manhattan and Brooklyn one city in reality as well as on paper. GCA members Tully & DiNapoli, Inc., Poirier & McLane, and Post & McCord were involved in building the Triborough Bridge.[5]

A World-Class Airport

Preparing for the World's Fair, the city filled another gap in its infrastructure, building its first world-class airport. Having served as an aviator during World War I, LaGuardia understood the importance of air travel. He was appalled that the city's only airport, Floyd Bennett Field on Barren Island, was hopelessly inadequate for the world's greatest city since its site was too foggy too much of the time to allow effective flight operations. With funds from the federal Works Progress Administration, the city built a large and modern airport on 550 acres of marshland at the old North Beach field in Queens. When it was completed in 1939, in time for the World's Fair, the facility included runways more than a mile long, four huge hangars, the most powerful searchlight in the world, and a large office space that later became the world's first airline lounge, the LaGuardia Admirals Club. The airport could accommodate 200 flights a day. Eventually, it would be named after the aviation enthusiast who envisioned it—LaGuardia Airport.

Foreseeing that even this facility would not meet the metropolis' aviation needs, New York began work in 1942 on another airport that also rose on a marshy shore front site in Queens. Originally called Idlewild, the airport would be named after President John F. Kennedy, after his untimely death in 1963.

Building a Network of Highways

Meanwhile, New York built the network of highways that enabled truckers and motorists to traverse the city, from the Bronx to Coney Island and from the West Side of Manhattan to the outskirts of Queens. In Queens, the Interboro Parkway and the Grand Central Parkway and its extension made it possible to drive from the other boroughs through Queens and out to Long Island's Nassau and Suffolk Counties Still other new highways were developed during this period in Queens, including the Northern State Parkway, the Whitestone Expressway, and the Cross Island Parkway.

Begun in 1934, the Belt Parkway forms a complete circle around Brooklyn and Queens, while the Gowanus Expressway goes west from downtown Brooklyn to the borough's southwestern tip.[7] Construction on the Brooklyn-Queens Expressway began in 1937, while work on the Long Island Expressway started in 1939.

Meanwhile, in Manhattan, a virtually complete belt was built around the island, including the West Side Highway, the Henry Hudson Parkway, and the East River Drive (later known as the Franklin D. Roosevelt Drive). GCA members James Stewart & Co. and Poirier and McLane helped to build the West Side Highway, while Mason & Hanger worked on the pile foundations.[8]

In the Bronx, the Mosholu, Pelham, and Hutchinson River Parkways all were constructed from 1935 through 1941, allowing cars and trucks to travel from the borough's southernmost points near Manhattan to its northern limits near Westchester County.

Linking Queens, Manhattan, and New Jersey

Together with bridges and highways, new tunnels were part of New York's expanded transportation network. During the 1930s and 1940s, the Port Authority of New York and New Jersey, built the Lincoln Tunnel linking Manhattan and the Garden State. With its south tube begun in 1934 and completed in 1937 and its north tube started in 1937 and finished in 1945, the Lincoln Tunnel stretches 1.5 miles under the Hudson River from the west side of Manhattan to Weehawken, New Jersey. GCA members Mason & Hanger, George J. Atwell Foundation Corporation, Arthur A. Johnson Corp., MacLean-Grove and Co., and Walsh Construction Co. constructed the Lincoln Tunnel.[9]   Meanwhile, Moses's Triborough Bridge and Tunnel Authority built the Queens Midtown Tunnel, connecting Queens with the east side of Manhattan between the major crosstown thoroughfares of East 34th and East 42nd Streets.[10]

Belt Parkway (left)
Belt Parkway riveters riveting floor beams of Spring Creek Bridge. March 28, 1940. Photographer unknown.
© New York City Department of Parks.

Belt Parkway Ribbon Cutting Ceremony (right)
Opening of the Belt Parkway. June 29, 1940.
Courtesy LaGuardia Archives.

Waste Management

As the world's greatest city, New York also generates the world's largest volume of waste. Until the end of the nineteenth century, the city government made little effort to clean the streets or dispose of garbage in ways that protected the health of the residents and the cleanliness of the air and water supply. Well-to-do families and prosperous businesses paid to have their trash removed, burned, or eliminated in other ways. In poor neighborhoods, far from being "paved with gold," the streets were covered with waste of all kinds, creating a stench and spreading disease.

While twenty-first century environmentalists complain about the consequences of Americans' love affair with the automobile, the means of transportation favored for most of the nineteenth century—the horse—created more primitive forms of pollution. By some estimates, by the middle of the nineteenth century, New York City's horses deposited 500,000 pounds of manure and 45,000 gallons of urine on the city's streets every day.[11] The horses' estimated life spans were only two and a half years, and, in a typical year—1880—an estimated 15,000 dead horses had to be removed from the city streets and their bodies disposed of somehow.

In 1881, the city founded its first Department of Sanitation (originally called the Department of Street Cleaning). At last, the city conducted regular garbage pickups. But the city's sanitation services still fell short of modern standards. "Putrescible" waste—the wet, smelly stuff that was most likely to rot—was carried by barge to be turned into fertilizer and other products in a "reduction plant" in Jamaica Bay. Coal ash and other household refuse from "ash cans" was dumped in landfills throughout the city.

These landfills kept growing well into the twentieth century. For instance, "Fishhook" McCarthy's smoldering dump in Corona, Queens, was immortalized as "the valley of ashes" in F. Scott Fitzgerald's novel, *The Great Gatsby*. Fitzgerald described it as "a fantastic farm, where ashes grew like wheat into ridges and hills and grotesque gardens."

Remarkably—and bravely—New York City decided during the depths of the Depression to invest in preserving its natural resources and managing the prodigious amounts of waste that the great metropolis produced. While resources were scarce, the needs were great. For decades, New Yorkers and their neighbors in other states had complained about the deteriorating condition of the surface waters surrounding the city. Barges operated by the city's Department of Street Cleaning dumped garbage into the waterways, and, often, instead of sinking or drifting off to sea, the garbage floated and washed up on the shore. Untreated sewage from the city's sewer system flowed into the waters surrounding the city and either killed off the city's much loved oysters, other shellfish, and fish or made the seafood that survived too contaminated to eat. It had been decades since youngsters or anyone else for that matter could safely take a refreshing dip in the Hudson or East Rivers.

Drawing of Tallman's Island Sewage Treatment Plant (above)

Rendering of plant's future development.
Courtesy of Municipal Archive Collection.

Horrace Harding Boulevard Sewer (opposite page)

Construction of double deck 3-barrel storm sewer along Horrace Harding Boulevard in Queens to serve the 1939 World's Fair. Construction work performed by GCA member Nicolas DiMenna & Sons, April 1938. Courtesy Municipal Archive Collection.

East River Drive (below)

Contract No. 7, from East 54th Street to East 64th Streets. September 19, 1939. Many GCA members were involved in the construction of the East River Drive including GCA member Nicolas DiMenna & Sons. Courtesy New York City Department of Parks.

2581 A6   4·7·38

Project 2 Contract 3 N.from Horace Harding B

Bowery Bay Sewage Treatment Plant (above)
Courtesy Municipal Archive Collection.

Coney Island Sewage Treatment Plant (above)
Courtesy Municipal Archive Collection.

The problem became impossible to ignore after 1934, when New Jersey prevailed upon the U.S. Supreme Court to enforce a 1888 statute against ocean dumping. After the ruling, New York simply tossed its garbage in open dumps that were scattered across the city like protuberant pox, stinking, smoking, and rat-infested. Most New Yorkers agreed that it was time—indeed, well past time—to find a better way to manage their rampant wastes.[12]

In 1935, with support from the federal Public Works Administration (PWA), New York City resumed work on its first modern sewage treatment plant on Wards Island, near the meeting point of the East River and the Harlem River. Occupying 52 acres of land and purifying 180 million gallons of water every day, the facility used the new technologies of chemical precipitation and sludge digestion. Also with support from the PWA, the city built a smaller modern plant on Coney Island in Brooklyn. Over the next 10 years, with assistance from the PWA, the city opened other state-of-the-art sewage treatment plants on Bowery Bay, Tallman Island, and in Jamaica, Queens. Meanwhile, in order to dispose of solid wastes by burning them, the city opened new incinerators at West 56th Street in Manhattan and in Flushing, Queens. The modern metropolis had found modern ways to manage its wastes.

Winning the War, Preparing for Peace

With America's entry into World War II in December 1941, new nonessential construction projects were put on hold. Appointed head of the national War Production Board's local salvage effort, Mayor LaGuardia threw himself and his city into the effort to secure the home front. Within days after the attack on Pearl Harbor, the city's construction contractors and building trades unions organized emergency crews prepared to make rapid repairs of buildings, water or gas mains, or electric facilities that might be knocked out in an air raid.

New York City's construction contractors pitched in wholeheartedly to help the war effort. The Rogers Corporation built a new Navy Pier on Staten Island, not far from the piers that had been built for World War I.[13] Another leading contractor, Underpinning and Foundation, upgraded the Floyd Bennett and Roosevelt airfields for the Army's use and built dry docks for the Navy at Perth Amboy, New Jersey.[14] Other contractors built ports, airstrips, and other essential facilities elsewhere in the United States, as well as in the European and Asian theaters of war.

People at the Opening of LaGuardia Airport (left)
Crowds at the opening of LaGuardia Airport in New York. December 6th, 1939.
© Bettmann/CORBIS

Grand Central Parkway (below)
View south along center line of Ditmars Avenue sewer from center line of parkway showing cut and foundation piles. October 10, 1935. Photographer unknown. Courtesy MTA Bridges and Tunnels Special Archive.

With victory in Europe on May 8, 1945, and victory over Japan on August 14 of that year, the Allies had defeated Nazism and fascism. Soon, the returning servicemen and women would begin the work of building the postwar prosperity, creating the world's first mass middle class, and opening new opportunities for those who had been excluded from the promise of American life. By making bold decisions and wise investments during the difficult years of 1930 through 1945, New York City built much of the physical structure that made possible the great progress of the decades to come, locally, nationally, and even globally. New York City's construction contractors and building trades workers did their part during the hard times, and they would continue to shape the future during the good times that lay ahead.

Foundations: Pushing Up

Why won't the Leaning Tower of Pisa stand up straight? Because it was built on top of a shallow foundation in unstable subsoil.

When New York's contractors started building towering office buildings—the famous skyscrapers—they faced similar challenges to those that had bewildered the builders in Pisa eight centuries earlier. After all, Manhattan is an island, and its soil is often weak and waterlogged. Just as they did with other difficulties that they faced, New York's builders developed new techniques for building strong foundations—technologies that changed the ways that construction was carried out throughout the nation and the world.

For builders, finding a firm foundation is a familiar problem. Bedrock might lie too far below the surface. Subsoil conditions might make it difficult to excavate a conventional basement style foundation. Or the structure might have to be built over water. Under such conditions, builders would connect the building and the bedrock with piles. Before the twentieth century, these piles were usually made of wood. For the past century, they've mostly been made of concrete, steel, or steel-reinforced concrete.

Sinking piles through layers of soil for hundreds of feet requires the exertion of force—lots of force. During the nineteenth century, the force was supplied by dropping heavy weights on top of the piles to drive them into the ground. Sometimes, mules or steam engines would hoist the weights to the tops of the derricks from which they were dropped. By the turn of the twentieth century, the gravity drop sometimes was replaced with repeating charges supplied by steam or other energy sources.

Riverside Drive Viaduct (above)
Construction of the foundation for Riverside Drive Viaduct between 155th and 161st Streets, looking north, 1928. Contractor, GCA member, Nicholas DiMenna & Sons. Courtesy Municipal Archive Collection.

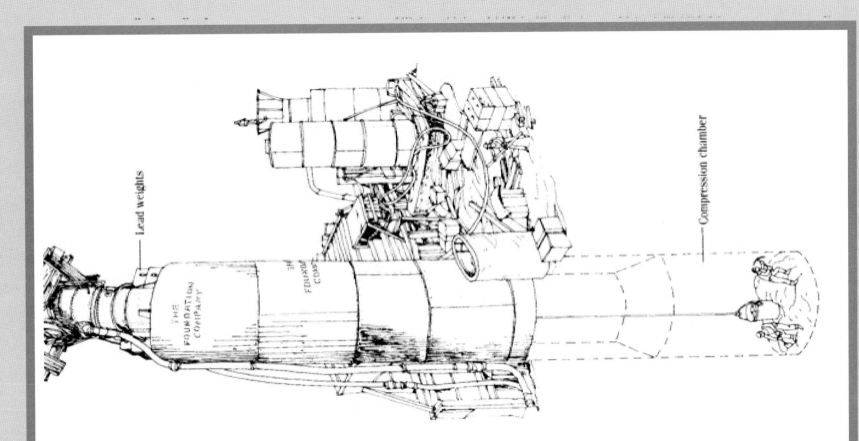

Lead weights

Compression chamber

To support a huge building, heavy duty pilings are required. When the Woolworth Company began building a new 58-story corporate headquarters in lower Manhattan in 1910, the architects and contractors faced the challenge of how to support its height and weight. Called "the Cathedral of Commerce," it would be the world's largest building from its opening in 1913 until it was surpassed in 1930 by the Chrysler Building, also in Manhattan.

The challenge's dimensions were daunting: The Woolworth Building would be 750 feet tall. Its weight of 223,000 tons—if spread evenly over the 30,000 square-foot site—would put a load of nearly 7½ tons on every square foot of the lot's surface. Below this surface was a layer of hardpan—a stratum of dense soil. And then, before the bedrock, was 105 to 121 feet of sand, the last 68 to 84 feet of which was saturated with groundwater. Even if it had been possible to pump out enough water to keep the excavated foundation pit from flooding, the loss of water from the surrounding soil would have had disastrous consequences. Adjacent buildings, streets, and miles of water, sewer, electric, gas, steam, and telephone lines—all would have collapsed.

What solution did the contractor—the Foundation Company (a GCA member)—devise? Sixty-nine concrete piers, up to 19 feet in diameter, would be extended down to bedrock. But how would the builders dig shafts that would allow the pre-made hollow piers to sink over a hundred feet below the surface through over 80 feet of water? The company's laborers had to climb down the shafts, passing through a cylindrical steel air lock to a pressurized chamber whose open bottom rested on the ground.

East River Drive (above)

East River Drive construction. Sinking of caisson. June 28, 1940.
© New York Department of Parks.

Hudson River Vehicular Tunnel (above, right)

Holland Tunnel construction. Rendering showing operations of sinking Hudson River Shaft Caisson. Rendering includes sand-hogs at bottom of shaft. Circa 1927.
Courtesy of Municipal Archive Collection.

Working under 40 pounds of pressure per square inch, the men shoveled the muck into buckets that were pulled to the top of the shaft. As they dug, the weight of the pier kept the bottom of the shaft resting on the ground. When they finished each shaft, the air lock was removed and concrete poured to fill the hollow center. Anchored to the top of each pier, so that it would sink with it as the pier descended through the sand, was a 40-foot-high sheet-steel cofferdam to keep the water out of the space in which the building's steel columns would be set. When the columns were in place, another sheet-steel cofferdam was driven around the perimeter of the lot and braced to prevent the collapse of adjacent streets and buildings while the 40-foot sub-basement was dug. The million dollar job took 10 months to complete.

Fifty years later, when the Underpinning & Foundation Company, a GCA member, built the foundation for 250 Broadway, a 29-story building across the street, they used quite a different method for sinking piles. After excavating 1,350 truckloads of subsoil to create a 33-foot-deep basement, underpinning used giant pile drivers to drive 860 steel-pipe pilings—about 16 miles worth—to a depth of 93- to 107-feet below grade.

Half a century separated the work on the Woolworth Building and 250 Broadway. But the common element was the ingenuity that enabled New York's contractors to devise new techniques for building foundations.

"Dewatering" in an Island City

New York City sits on top of—and is surrounded by—water. Built on three islands and one peninsula, New York is nestled amidst a patchwork of streams and estuaries, freshwater wetlands, and tidal marshes. Moreover, much of the city's shore front s formed from landfill. Thus the water table often lies a short distance from the surface.

Having water so near ground level poses a special challenge to New York's construction industry. Building foundations, tunnels, and other underground structures requires a dry work environment. But water, whether above ground or below, will always flow downhill. So any hole dug for a foundation, tunnel, or utility line that reaches below the water table will fill with water. In order to excavate with any efficiency, or to construct permanent structures in such holes, this water must either be prevented from entering the site or be removed from it.

The process of eliminating excess water from construction sites is known as "dewatering." Over the past 100 years, New York's construction industry has exhibited an extraordinary amount cf ingenuity in developing new techniques for dewatering the sites for skyscrapers, subways, tunnels, and other projects.

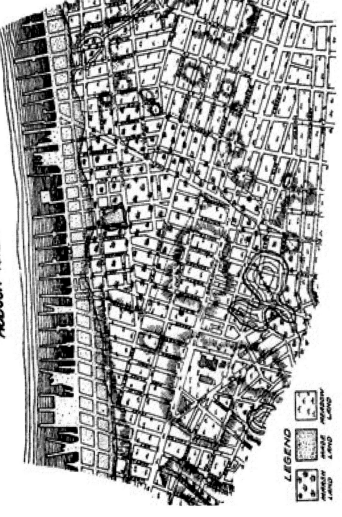

Lower Manhattan (above)
Map of a portion of lower Manhattan drawn by Townsend Macoun, 1783.

Despite the fact that people have been building structures beneath the water table for many millennia, the science of dewatering is relatively new. Before the beginning of the twentieth century, builders usually used primary pumps (or caissons with compressed air) to remove the water from underneath construction sites.

During the twentieth century, in addition to caissons and primitive pumps, builders used three basic methods—often in combination with each other—to keep water out of excavation sites with flooding or flows of groundwater.

The first method is to build a structure, such as a drainage ditch or a "cofferdam" that prevents water from entering a site. The second method is to remove water from a site by pumping it out. The third method keeps water out by making the ground underneath the excavation site less porous or permeable. This can be done by injecting materials into the ground such as cement "grout" (a mixture of water, sand, and cement that hardens like mortar). In a major technological advance, the construction industry devised a new way of keeping water out—freezing the site with liquid nitrogen or refrigerated brine to keep it stable during the time of construction.

Over the years, all three of these methods have become more sophisticated. For instance, cofferdams are often constructed by driving sheets of corrugated steel into the ground in such a way that the edges of adjacent sheets overlap or interlock to form a continuous wall.

Manufacturers Hanover Trust Building (above)

Two views of the cofferdam foundation for the 22-story Manufacturers Hanover Trust Building at Broad and the aptly named Water Street, in 1967. After digging a ditch around the site perimeter, the contractor drove 600 sheets of piling capable of withstanding water pressures up to 1,600 lbs/sq. ft. down to hardpan. Although the water table was only 5 feet below the sidewalk—and the excavation went 30 feet below that—the site stayed dry. Courtesy General Contractors Association of New York.

World Trade Center (opposite page)

GCA member Moretrench operating from a central ejector pump station is dewatering the area adjacent to the PATH tubes. The soil consists of silt, muck, and very fine sand. Courtesy Moretrench American Corp.

Similarly, contractors have developed new ways of pumping water out of excavation sites. One technique—the "wellpoint method" was first used on Sir Robert Napier's march to the city of Magdala during Great Britain's invasion of Ethiopia (then known as Abyssinia) in 1896. Each "Abyssinian tube"—a short, narrow-diameter pipe—was driven deep into the ground with a sledgehammer. If it struck water, a traditional village-type hand pump drew it out.

Starting in 1901, such tubes were used in North America. But the first significant improvement in their use came in 1925, when Thomas Moore, the founder of the Moretrench Company (a GCA member), made an important discovery. If a jet of water was sprayed through the inner tube as it was pushed into the ground, the water bubbling back out of the hole produced an opening around the outer tube into which sand could be shoveled. After the sand was put in place, the outer hole-punching tubes were withdrawn from the well, leaving behind the smaller pipes through which the water could be pumped from the hole. Moore found that the sand functioned as a readily permeable filter through which the water could be suctioned and removed from the construction site. In fact, the tubes could be driven in a linked series so that one pump could pull water from every wellpoint. Since underground water in the area surrounding the hole could be pulled to the bottom of the wells by gravity (and the force of the suction pump), the water table in the whole area surrounding the wells would be lowered so that excavation could take place without breaching the waterline.

The Moretrench Company continues to provide specialized dewatering services on complicated dewatering projects around the world.

During the decade beginning in 1910, the Catskill aqueduct system, which brings water into New York City, was being constructed. Since then, tunnel workers have pumped high-pressure grout through nipple-covered holes in a tunnel's steel walls to stabilize the ground underneath and make it less permeable. Sometimes more sophisticated stabilizing techniques are needed to address more complicated problems. For instance, in many parts of the city, layers of fine-grained subsoil materials, such as sand and silt, are so saturated by flowing groundwater that they are called "running sand," "quicksand," or "bull's liver."

Responding to these challenges, builders keep developing more technologically advanced methods of keeping water out of excavation sites. In a joint venture, three GCA contractors—the Foundation Company, George M. Brewster & Sons, and Joseph Miele Contracting Co.—dug the foundation for the Chase Manhattan Bank's new 60-story headquarters at William and Pine Streets, near Wall Street, in lower Manhattan. At 60-feet below street level, workers ran into waterlogged subsoil. Although the workers had already driven support piles and used pneumatic caissons to remove six stories' worth of sand, the quick-flowing, highly unstable quicksand made further progress impossible.

East Side Access (left)
GCA member Moretrench used sodium silicate permeation grouting to alleviate water infiltration to a work shaft for the East Side Access project. Groundwater was present at 20 feet. This shaft was excavated through granular overturden to rock at depths between 60 and 80 feet below working grade. Courtesy Moretrench American Corp.

1964 Moretrench ad (above)
Courtesy General Contractors Association of New York.

Their solution? They drove 200 pipes—each only 1 5/8 inches in diameter—to depths of 45 to 75 feet into the ground. Then they pumped 90 tons of calcium chloride and 45,000 gallons of silicate of soda into these wells at pressures of up to 600 pounds per square inch. When exposed to each other, these chemical compounds produce a hydrated calcium silicate, which is a primary ingredient in cement. When hardened, this substance behaves much like limestone—and halts the flow of water. This was the first use of chemical soil stabilization in New York City.

In a subway in lower Manhattan during the late 1970s and early 1980s, T. Moriarty and Son developed an innovative dewatering plan. GCA President James Moriarty, Jr., who now heads the family firm, recalls "In one of those projects, they wanted to dewater a problem area of the tracks, where they had a significant infiltration of water. The plan on that job called for it to be done with a series of small wells that would have been drilled from within the track envelope. The contract called for us to use a smaller rig in the tunnel which we felt could take a significantly longer time to work. We proposed to the New York City Transit Authority that we open the street from a certain distance apart and drill larger wells so that we could keep the water down and prevent the water infiltration. We actually had to make the opening from the street and then drill the well from the street with the track outage so the trains were not moving past. At the completion of the job, we closed up the street from above. And it did work. That is the process that is generally used now."

For New York City contractors digging into unstable waterlogged subsoil, another technologically advanced technique is freezing by liquid nitrogen. This method was first used in New York City by GCA member Poirier & McLane to bore an interceptor sewer tunnel to the Newtown Creek sewage treatment plant in Brooklyn.

One of the most famous dewatering challenges in recent decades was the "bathtub" over which the World Trade Center towers were built. Because of a high water table on this landfill, in order to keep the water from filling the excavation site, a cutoff wall had to be sunk around the perimeter of the area before it was excavated. To do this, the contractor, ICANDA, Ltd. (a GCA member), used rigs equipped with clamshell buckets to dig a 3-foot-wide, 70-foot-deep trench around the 3,400-foot perimeter of the site. To keep the walls of the trench from caving in before the cement was poured, they pumped a mixture of bentonite and water into each 22-foot section of the trench as soon as the dirt was removed. (Bentonite, a volcanic clay ash, absorbs great volumes of water to produce a gel-like slurry that is strong enough to withstand cave-ins, yet light enough to be displaced when concrete is poured into it.)

Prefabricated steel frames were then inserted into the bentonite slurry, and concrete was pumped in. To maintain stability while the wall was being built, the 22-foot sections were dug in alternation so that undug sections remained between each two concreted sections. Liquid grout was injected under high pressure into the soil behind the completed sections of concrete. When the grout had hardened, diagonal shafts were drilled down through the walls and into the grout so that tieback cables to anchor the walls in place could be run through the shafts and themselves grouted into place. When the walls and tiebacks were completed on both ends of an undug section, it was then dug, and, when the wall section was completed, it was joined to those on either side with an interlocking joint.

This was so well planned and completed that the site remained dry, excavation was conducted, and the towering World Trade Center buildings were constructed. But the best tribute to the dewatering came decades later, under tragic circumstances. After the twin towers collapsed during the terrorist attacks of September 11, 2001, the underground walls held, what had been the "bathtub" was not filled with water, and lower Manhattan was not flooded.

World Trade Center Construction (above)
Construction of slurry wall for World Trade Center.
Courtesy Port Authority of New York and New Jersey.

Timeline: 1946–1962

1946	1947	1948	1949	1950	1951	1952	1953	1954
Tunnel Authority and Triborough Bridge Authority Merge	Erie Basin Whitestone Interceptor Sewer Tunnel	Idlewild Airport Opens	East Bronx Intercepting Sewer	NYS Thruway Authority established	Wards Island Pedestrian Bridge	Hunts Point Sewage Treatment Plant	Unionport Bridge	Pulaski Bridge
Mayor William O'Dwyer		Fresh Kills Landfill Opens		Sewer charge instituted (Local Law 67)	Chelsea Pump Station to pump Hudson River water into Delaware Aqueduct	Rockaway Sewage Treatment Plant	Bruckner Boulevard Bridge (formerly Eastern Boulevard Bridge)	New York Central Bridge over Harlem at 133rd Street
				Mayor Vincent R. Impellitteri	First Natural Gas from Texas arrives via the Transcontinental pipeline across Hudson at 134th Street. The 1,840 mile "Big Inch," whose construction started 2 years before will allow ConEd and Brooklyn Union Gas to stop manufacturing gas from coal	Owl's Head Sewage Treatment Plant	Port Richmond Sewage Treatment Plant	South Shore Incinerator
				Betts Avenue Incinerator		New York City Transit Authority	Gansevoort Incinerator	North Shore Marine Transfer Station
				Brooklyn-Battery Tunnel opens		Transcontinental Gas Pipe Line Corporation's Narrows Tunnel	Mayor Robert F. Wagner	West 216th Street Marine Transfer Station
				Rondout Reservoir (Delaware System)				Neversink Reservoir (Delaware System)
								Pepacton Reservoir

1955	1956	1957	1958	1959	1960	1961	1962
Roosevelt Island Bridge	Heliport 30th Street	Lincoln Tunnel	Lemon Creek Bridge Opens	Arthur Kill Railroad Lift Bridge Opens	Morgan Avenue Interceptor Sewer Completed	Throgs Neck Bridge	Broadway Bridge
Third Avenue El	Park Avenue Bridge Opens	The City's Last Trolley Car (Queensboro Bridge Railway Co.) makes its last run	New England Thruway	Brooklyn Union Gas acquires Brooklyn Borough Gas Co.	Greenpoint Avenue Interceptor Sewer Completed	Throgs Neck Expressway	George Washington Bridge Double-Deck
Queens-Midtown Expressway	Highway Act of 1956	East 73rd Street Incinerator	Horace Harding Expressway (now Long Island Expressway)	Con Ed's Travis Plant starts new generating unit	Kent Avenue Interceptor Sewer	Southwest Brooklyn Incinerator	Prospect Expressway
Oakwood Beach Interceptor Sewer	Whitestone Parkway	Brooklyn Union Gas acquires the New York and Richmond Gas Co. and Kings County Lighting Co.	Johnson Avenue Interceptor Sewer		Freezing Subsoil	Southwest Brooklyn Marine Transfer Station	Sheridan Expressway
West 135th Street Marine Transfer Station	Major Deegan Expressway		Hunts Point Marine Transfer Station		Chemical Treatment of Subsoil	College Point Incinerator	Bruckner Expressway
Gansevoort Marine Transfer Station	Oakwood Beach Sewage Treatment Plant Opens		Hamilton Avenue Incinerator				Con Ed's Indian Point 1 Nuclear Power Plant
	Rector Street Interceptor Sewer completed		Greenpoint Incinerator				
	Greenport Marine Transfer Station		Transcontinental Gas Pipe Line Corporation's third underwater pipeline				

"Everywhere I go, I see buildings that our company helped to build or repair—Lincoln Center, the Chrysler Building, the Hilton Hotel, Citicorp."

Theodore Civetta,
President & Secretary, John Civetta & Sons, Inc.

1946–1962:
Capital of the World

At 7:01 p.m. on August 14, 1945, New Yorkers learned that the Japanese had officially surrendered, ending combat in World War II. As Americans rejoiced in V-J Day, more than 2 million people flocked to Times Square for a spontaneous celebration that lasted long past midnight. For the first time since the attack on Pearl Harbor, the neon lights were shining once more on the towers, theaters, and restaurants. The evening's exuberance was captured in the famous photograph, "the Kiss," which showed a sailor, in his uniform, embracing a nurse, in her uniform. The war was over, and people were partying as if it were New Year's Eve.[1]

It wasn't the dawning of a new year, but it was the beginning of a new era—the postwar era. After 16 years of suffering and sacrifice—first, the Great Depression and, then, the Second World War—people were eager to return to peace and prosperity. But most Americans were also reluctant to repeat the mistakes made in the aftermath of World War I, when the nation shrunk from its challenges at home and abroad, refusing to join the League of Nations and retreating to economic policies that predated the Progressive Era. This time, the United States assumed the leadership of the world's democracies and invested in educa-

tion, housing, transportation, and other urgent needs. While the Cold War, the Korean War, and the Vietnam War would take a terrible toll, the wise decisions that the nation made in the years following World War II paved the way for shared prosperity in this country and the spread of democracy throughout the world.

Historic Highpoint, Headquarters City

For New York City, the two decades after World War II would be a historic high point. The city emerged from the war as the center of the nation's economic and cultural life, and, in the years ahead, New York firmly established its leading role not only nationally but also internationally. As the essayist E. B. White wrote, while New York City was not the governmental center of the nation or even the state, it was "becoming the capital of the world."

After World War I, the United States decided not to join the League of Nations, even though the international organization had been the brainchild of President Woodrow Wilson. After World War II, the United States not only joined but hosted and helped to lead the new world parliament, the United Nations.

The Kiss (above)
A jubilant American sailor clutching a white-uniformed nurse in a joyful, back-bending, passionate kiss while thousands jam the Times Square area to celebrate the long awaited victory over Japan. Photo by Alfred Eisenstaedt//Time Life Pictures/Getty Images

Van Wyck Expressway paving (opposite page)
Camera at 133rd Avenue Bridge, facing south. June 26, 1951. Photographer: Roy Foody for New York State Department of Public Works. Courtesy MTA Bridges and Tunnels Special Archive.

In 1946, the U.N.'s first Secretary-General, Trygve Lie, declared that he hoped that its headquarters would be in New York City. Then the question became: in such a densely settled city, where was there a desirable and accessible site for the institution that embodied humanity's hopes? The diplomat, philanthropist, and future governor, Nelson A. Rockefeller, had an answer: a 4-block, 17-acre tract of land in midtown Manhattan, along the East River. Ironically, for the future site of an organization dedicated to world peace, the area was filled with slaughterhouses and cattle pens and was known as "Blood Alley." Another problem was that the site had recently been bought by the developer William Zeckendorf who had his own plans for the area.

Undeterred, the Rockefeller family bought the land for $8.5 million and donated it to the U.N. With an interest-free loan of $65 million from the U.S. government, the U.N. began designing and building its new headquarters.[2] Designed by the famed Swiss architect Le Corbusier, the 38-story headquarters building, with its modernistic glass and steel structure, became a modernist icon and a model for the new corporate headquarters that were springing up throughout Manhattan. GCA members, George Fuller, Walsh Construction, and Slattery Contracting Co., were involved in constructing the U.N. building.[3]

By the early 1950s, New York was becoming the headquarters city not only for the United Nations but also for hundreds of leading national and multinational corporations. Of the country's 500 largest manufacturing companies, at least 135 were headquartered in New York City, including GE, IBM, RCA, Standard Oil, Union Carbide, and U.S. Steel, to name only a few. Meanwhile, New York was even more dominant in advertising, the media, and related industries. For instance, the three major television networks—ABC, CBS, and NBC—were headquartered in New York, as were the leading news magazines, Time and Newsweek. Wall Street, of course, remained the world's financial center, and, for the nation's banks, brokerage houses, and other financial institutions, Manhattan was the location of choice.

During the years immediately after World War II, New York City would be a center of blue-collar and white-collar jobs. With thousands of factories employing more than a million workers, New York was the nation's leading manufacturing city. In fact, Manhattan south of 42nd Street employed more industrial workers, per square mile, than any other city in the world. To be sure, New York specialized in garments, electronics, plastics, toys, jewelry, and other industries with relatively small companies and factories, not basic industries such as steel and automobiles. New York was also the world's biggest port, handling 150 million tons of cargo every year.

United Nations Building (above)
Construction of permanent headquarter's of the United Nations in New York City by joint venture of George Fuller, Co., Turner Construction, Walsh Construction, and Slattery Contracting, 1949. Courtesy Skanska USA Civil Northeast Inc.

**Battery Park Seawall
(opposite page)**
Construction of Battery Park Seawall by GCA member Spearin Preston & Burrows. Looking east at bulkhead wall north boat basin. November 13, 1946. Courtesy Spearin Preston & Burrows.

Of course, construction would also be a major source of unionized, good-paying, blue-collar jobs, as the city, state, and federal government and private developers built new housing, office buildings, bridges, highways, subways, aqueducts, sewage treatment plants, and other structures. During the postwar era, the construction industry took the lead as New York and the nation enjoyed a shared prosperity and built a mass middle class and a physical infrastructure that made this progress possible.

Back to Work Building New York City

With World War II won, New York City went back to work building, rebuilding, and repairing the physical structures of the world's greatest city. With the wear and tear resulting from transporting troops and munitions during the war and with deferred maintenance dating back to the beginning of the war or even the onset of the Great Depression, New York City had a huge job to do.

Meanwhile, national and international developments influenced the extent to which New York City received the resources it needed for construction projects. The nation's economic policies had shifted dramatically from the penny-pinching laissez-faire of the 1920s

to the Keynesian pump-priming of the 1930s and 1940s. Once victory in World War II appeared assured, policymakers feared that the economy would slow down with reductions in military spending. Returning veterans would join the ranks of the unemployed, and the nation would once again plunge into a recession, just as it had after World War I. Therefore, Presidents Franklin D. Roosevelt and Harry S. Truman and the Congress created programs from the G.I. Bill of 1944 to the Housing Act of 1949 that invested in education, housing, and other needs; pumped money into the economy; and made sure that veterans could attend college as well as rejoin the labor force.

Soon, it became clear that the postwar era would not be entirely peaceful, military spending would remain at record levels, and there would not be unlimited federal funding for purely domestic programs. With the establishment of communist regimes in Eastern Europe, the Soviet Union's acquisition of nuclear weapons, and rising tensions between the two super powers, the world was mired in the "Cold War." The next four decades were marked by an arms race, ideological struggles, fierce rivalries for the allegiance of uncommitted nations, standoffs in Cuba and Berlin, and wars in Korea, Vietnam, and Afghanistan.

Construction of Rockefeller Center (above)
Rockefeller Center under construction, 1932.
Courtesy Museum of the City of New York, The Wurts Collection.

But, thankfully, there was not a global nuclear conflict that could have annihilated much of humanity, poisoned the planet, and destroyed modern civilization. Instead, there was a mostly peaceful competition in arenas ranging from the development of missile systems to the exploration of space as well as the pursuit of economic, educational, and scientific progress that would prove one system's superiority over the other. On the federal, state, and local levels, those who promoted public investments, from education to transportation, made sure to present their programs in terms of national security as well as social needs. New York's public officials, particularly the ever-present Robert Moses, now the city's construction coordinator, adapted to the imperatives of the Cold War.

New York City had far more competing demands for new facilities than the federal, state, and local governments could afford to fund, even with all the project revenue bonds. New public schools, public hospitals, police and fire stations, sewage treatment plants and incinerators, highways, bridges, tunnels, and subway lines—all had to be built, while other public facilities had to be rebuilt or repaired. But, in addition to the conflicting claims on public funds, many of the engineers in New York City government had left for more lucrative jobs in the private sector, and there was also a shortage of skilled construction craft workers.

As the dominant figure in New York's construction programs, Moses was able to set priorities for the city's projects. His most important priorities included: the three elevated expressways that he had proposed across lower, midtown, and upper Manhattan; $100 million worth of new parking facilities; and a new coliseum in midtown Manhattan. Attuned to the political realities of the times, Moses presented many of these projects as bolstering the city's defenses.

Within a year after the outbreak of the Korean War, many federal resources were devoted to defense industries that tended to be located far from New York. In 1956, Congress passed and President Dwight D. Eisenhower signed the Highway Act of 1956—in keeping with the Cold War spirit, its full name was "the National Interstate and Defense Highways Act." The federal government would provide 90 percent of the funds for highway projects, and state governments would pay the remaining 10 percent. Shrewdly, Moses was a leading advocate of the Highway Act, even winning a national essay contest sponsored by General Motors in support of the legislation.[4] Soon, the pace of construction, particularly highways, roads, bridges, and tunnels, picked up throughout the city.

Karl Koch Erecting 50th Anniversary (left) 1956. Courtesy Skanska Koch.

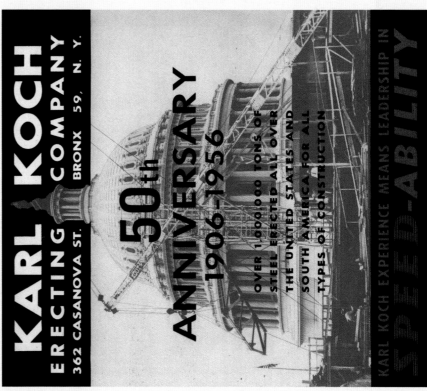

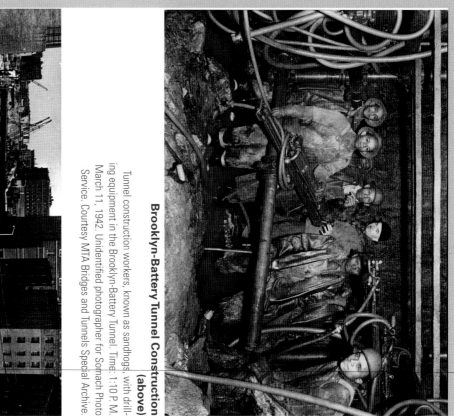

Brooklyn-Battery Tunnel Construction (above)

Tunnel construction workers, known as sandhogs, with drilling equipment in the Brooklyn-Battery Tunnel. Time: 1:10 P.M. March 11, 1942. Unidentified photographer for Somach Photo Service. Courtesy MTA Bridges and Tunnels Special Archive.

Cross Bronx Expressway (above)

Cross Bronx Expressway construction. No date. Photographer unknown. Courtesy MTA Bridges and Tunnels Special Archive.

Highway, Bridges, and Tunnels

In retrospect, the postwar era was a golden age for New York City's transportation systems. Ridership on the city's subways and buses reached peaks that had never been surpassed.[5] Although air travel had increased exponentially since the opening of the North Beach Airport in 1939, long-distance passenger trains still played an important part in intercity travel, and ocean liners still regularly plied the seas between New York and Europe. The freight railroads serving the city were hauling more cargo than ever, even though they were suffering from strict regulation, increased competition from trucks, and the costs of floating cars or lightering goods across New York Harbor.

The transition from horse-drawn vehicles to trucks and automobiles had long since been completed. The city was beginning to experience a level of traffic congestion that it had not endured since before its bridges, tunnels, highways, and subways were built. However, only 10 percent of the city's streets were badly snarled with traffic[6] —the term "gridlock" would not be coined for decades to come.[7] The flow on those constricted arteries was eased by new traffic-control techniques such as one-way streets and parking meters. Meanwhile, the newer highways, bridges, and tunnels were not nearly as crowded as they would later become.

Still, New York City needed new highways, bridges, and tunnels. Starting in the immediate aftermath of World War II, Moses pushed through the projects that had been delayed. One of the earliest—and most important—of these projects was the Brooklyn-Battery Tunnel. Construction resumed as soon as the federal War Production Board gave it the green light in November 1945. When the tunnel opened on May 25, 1950, it had a capacity of 16 million vehicles a year, and, at 9,117 feet from portal to portal, it was the longest underwater tunnel in the country.[8] Work also began on the Bronx River Parkway and the Brooklyn-Queens Expressway. Both had been begun well before World War II and were completed in 1952 and 1964, respectively.

Then the city began building new projects again. The Cross Bronx Expressway was begun in 1946 and completed in 1963. Next, work started on the Van Wyck Expressway in Queens, which was also completed in 1963, and the Harlem River Drive in Manhattan, which was completed in 1964. Within the next few years, the Major Deegan Expressway and the New England Thruway, both in the Bronx, got underway in 1949 and 1950 and were finished in 1956 and 1958, respectively. GCA members Del Balso Construction Corporation, Rusciano & Son Corp., Gull Contracting Co., Inc., Tully & DiNapoli, Inc., Johnson, Drake & Piper, Inc., Charles F. Vachris, Inc., and Slattery Contracting Co. were involved in building the Major Deegan Expressway.[9] GCA member Yonkers Contracting Co. helped build the New England Thruway.[10] In Brooklyn, the Prospect Expressway was begun in 1953 and completed in 1962, while the Long Island Expressway in Queens was started in 1954 and finished in 1958.

Van Wyck Expressway (above)
Aerial view of the Van Wyck Expressway Extension and Long Island Expressway interchange nearing completion. Pavilions for the 1964–1965 New York World's Fair are under construction at upper right. Photographer: Skyviews New York. December 12, 1963. Courtesy MTA Bridges and Tunnels Special Archive.

Verrazano-Narrows Bridge Cable Construction (above)
View looking southeast on the Staten Island Tower showing cable spinning wheel passing over tower saddle delivering wire for one of the bridge's four cables. July 10, 1963. Photographer: Paul Rubenstein, Lenox Studios Photography. Courtesy MTA Bridges and Tunnels Special Archive.

East River Drive (above)
View of East River Drive construction taken from the 15th floor of 120 Wall Street. Photographer unknown. August 10, 1953. Courtesy MTA Bridges and Tunnels Special Archive.

With the passage of the $33 billion Highway Act in 1956, Moses identified a billion dollars worth of new projects to recommend for federal funding on the grounds that they promoted interstate travel. In the Bronx, these projects included the Bruckner and Sheridan Expressways, begun in 1957 and 1958 and completed in 1962. In Queens, the Highway Act paid for the Clearview Expressway, which was begun in 1957 and completed in 1963. In Brooklyn, the federal funds helped to complete the Brooklyn-Queens and Prospect Expressways and upgraded the Gowanus Expressway. In Staten Island, the Highway Act funded the Clove Lakes and West Shore Expressways, which were both begun in 1959 and were completed in 1964 and 1966, respectively.

For Moses, bridges were even more important than highways. During the postwar era, his most ambitious project by far was the Verrazano-Narrows Bridge, connecting Brooklyn and Staten Island. With its huge towers soaring 650 feet above the water, its double-deck roadway suspended 226 feet high, and a span of 4,260 feet, it was the largest suspension bridge in the world and is still the

largest in the United States. Begun in 1959 and completed in 1964, it contributed to the growth of Staten Island and offered a new route to travel by car from New Jersey to Brooklyn, Queens, and Long Island. GCA members Slattery Contracting Co., Inc., J. Rich Steers, Inc., Frederick Snare Corporation, Arthur A. Johnson Corporation. Peter Kiewit Sons Company, and Anthony Grace & Sons, Inc., were involved in building the Verrazano-Narrows Bridge.

Before beginning the Verrazano-Narrows Bridge, the city began building other bridges, including the Roosevelt Island Bridge to Queens, which was started in 1952 and completed in 1955, the Throgs Neck Bridge between the Bronx and Queens, which was started in 1958 and finished in 1961, and the Alexander Hamilton Bridge between Manhattan and the Bronx, whose construction took from 1960 to 1963. GCA members Merritt-Chapman & Scott Corp., J. Rich Steers, Inc., Frederick Snare Corp., Fehlhaber Pile Co., Slattery Contracting Co., Inc., Steers-Snare, Horn Construction Co., Inc., and Queens Structure Corporation were involved in building the Throgs Neck Bridge.[11]

The United Nations (above)
Construction of the United Nations. The work was performed by Slattery Contracting.
Courtesy Skanska USA Civil Northeast.

Bronx-Whitestone Bridge (above)
View from Queens of the Bronx-Whitestone Bridge under construction with all but the center unit of the bridge floor in place. 1939. Photographer unknown. Courtesy MTA Bridges and Tunnels Special Archive.

225th Street Bridge (below)
1962. The work was performed by Slattery Contracting.
Courtesy Skanska USA Civil Northeast.

As it turned out, the Verrazano-Narrows Bridge was the last major bridge to be built in New York City. In the years ahead, the city would focus on maintaining—and, when necessary, replacing—bridges that had already been built.

Mass Transit

Unfortunately for New York's subway system and the straphangers who relied on it, the same public funds that subsidized highway construction were not available for mass transit. While the federal Highway Act paid for roads and bridges, it did not provide funding for the city's subways. Nor would the Port Authority or any of Moses's special authorities. In fact, Moses refused to allow the Verrazano-Narrows Bridge to be designed to allow future transit lines to cross it, just as the subways use the Manhattan Bridge.

However, Moses did use the Cold War concerns about national security when he sought federal funds for his favorite subway project. In yet another ingenious appeal, he argued that the federal government's new civil defense program, which paid half the costs of public bomb shelters, should consider several proposed underground projects as potential bomb shelters. Therefore, he maintained, the federal government should pay the costs of underground parking facilities and "strategically located" stations at 14th, 42nd, and 59th Streets on the yet-to-be-built Second Avenue subway. But, in spite of Moses's formidable powers of persuasion, the request was not favorably received. Congress declared that such "self-sustaining" facilities were ineligible for federal funding, but it did agree to pay half the cost of whatever additional "hardening," such as thicker concrete, or "softening," such as additional toilets, were needed to adapt such facilities for use as bomb shelters.

During the decade and a half after World War II, there was no new subway construction. But this was not for want of trying to build new subways. During World War II, when it was impossible to obtain the public funds, the building materials, or the skilled labor for subway construction, Mayor Fiorello LaGuardia promised that, after the war was won, the city would build the long-delayed line on Second Avenue on the east side of Manhattan.

Plans for the Second Avenue subway had been approved by the city's Board of Transportation in 1929—less than a month before the stock market crash that led to the Great Depression that made new subway construction unaffordable. Then came World War II. When the war ended and construction materials were available, there were still no funds for building a new subway line.

By 1951, there was widespread support for an entire new subway system that would include the Second Avenue line and would run from lower Manhattan to the Bronx, with connections to Brooklyn and Queens. In November, New York State voters approved a $500 million exemption from the city's debt limit, so that the Second Avenue line could be built.

In anticipation of the Second Avenue subway being built, the Second Avenue elevated line was torn down in 1942. In 1955, the last of the city's elevated lines—the Third Avenue el—was demolished. In 1957, the city's last trolley line, which had rumbled across the Queensboro Bridge for 48 years, was discontinued.

In an important addition to the city's transportation facilities, a new airport opened in 1948. It was built in Queens on about 1,000 acres of the old Idlewild Golf Course—thus, its name, Idlewild Airport. Until the completion of the international arrivals building in 1957, it had only one terminal. Starting in 1958, another 4,000 acres were added to the airport, and many of the leading international airlines built their own terminals. After the assassination of President John F. Kennedy in 1963, the airport was renamed the John F. Kennedy International Airport.

The Port and the Rail Lines

From their earliest days, the region's railroads and New York City's port were tightly intertwined. After all, any development that affected the railroads or the port would eventually affect the other—and would have an enormous impact on the economy of the entire metropolitan area. That was one reason why managing the relationship between the railroads and the port was one of the goals of the Port Authority, which was established in 1921.

Head of Sewer Workers in a Manhattan Sewer (above)
July 1948. Photographer: Jerry Cooke. © Jerry Cooke/CORBIS

Springfield Blvd. (inset)
Construction of Springfield Boulevard Sewer by GCA member Spearin Preston & Burrows. November 29, 1951. Courtesy Spearin, Preston & Burrows.

During the years after World War II, the railroads and the port were challenged by economic and technological changes. The war had strengthened the trucking industry, whose increasing dominance drew business away from ships and trains. Unable to withstand the onslaught of government-subsidized competition from trucks, the nation's eastern railroads continued their inexorable slide into bankruptcy.

Meanwhile, containerization was transforming the shipping industry. Shipping containers are standard metal boxes that can be transferred between ships, trucks, or trains, while their contents remain intact. In addition to almost entirely eliminating the need for stevedores, containerization benefited the New

Jersey side of the harbor, which was blessed with plenty of land and tracks. This reduced the appeal of the Manhattan and Brooklyn side, where land was scarce and rail access limited. Containerization originated with the Sea-Land Company, the first shipping company to use cargo containers.

Just as the railroads were sliding into bankruptcy, the New York piers also were declining. During World War II, the port's operations had reached their highest level ever. For several years after the war, the port remained the world's largest, handling half the nation's foreign trade. But, relative to its rivals on the eastern seaboard, the port's position was quickly slipping. For instance, in 1946, Boston's imports increased by 62 percent, and Baltimore's imports increased by 67 percent, but New York's decreased by 5 percent.[12]

For years, the city's public works agencies—the Port Authority, the World Trade Corporation, and the Department of Marine and Aviation—could not agree on how to revive the piers. In 1956, the Port Authority evicted the shipping business that had lined two miles of shore front between the Brooklyn Bridge and Erie Basin so that it could build 10 new piers there. While the evicted businesses took up quarters in new facilities on the New Jersey side of the harbor, the Port Authority spent $95 million on Brooklyn piers that were completed in 1963. But few of the evicted businesses moved back across the harbor to use the pier, nor did many new tenants emerge.

Sanitation

Leading the nation in population and economic activity, New York City generated growing amounts of solid, liquid, and gaseous waste. Disposing of this waste required the construction of sanitation facilities of many kinds.

Opened in 1948 on 4.6 square miles in Staten Island, the Fresh Kills landfill became one of the largest heaps of waste in the world. Soon, the landfill received an average of 20 barges—each carrying an average of 650 tons of garbage—every day.

Why did the bucolic borough of Staten Island agree to accept what could crudely be called a dumping ground? Moses promised the Staten Island borough president that the landfill would be closed to "raw" garbage by 1951. Enough incinerators would be built to ensure that all the city's garbage would be "cooked."

The city government did its best to keep the promise. From the late 1940s through the early 1960s, the Department of Public Works built incinerators at Betts and 53rd Avenues, Wortman Avenue and Forbell Street, and at College Point Causeway and Whitestone Expressway, all in Queens; at Gansevoort Street on the Hudson River in Manhattan; and on Hamilton Avenue and 16th Street, Apollo Street and Newtown Creek in Greenpoint and Bay 41st Street and 25th Avenue, both in Brooklyn. But old incinerators at East 73rd Street in Manhattan, College Point and Ravenswood in Queens, at Paerdergadt Basin in Brooklyn, and at Richmond Terrace in Staten Island all closed during the same period. The net gain in the city's incineration capacity was less than 3,000 tons a day less than about half the raw garbage that the city produced.

Construction of East 73rd Street Incinerator (left)
Courtesy Municipal Archive Collection.

Staten Island Anchorage, Narrows Bridge (below)
Construction by GCA
member Johnson-Kiewit.
October 2, 1962.
Courtesy General Contractors
Association of New York.

World Trade Center Construction (opposite page)
Steel erection for World Trade Center by GCA
member Karl Koch Erecting. Courtesy of Port
Authority of New York and New Jersey.

The city still needed to get raw waste to the Fresh Kills landfill. To get the garbage to Staten Island, the city built new or replacement marine transfer stations at 31st Avenue and Flushing Bay in Queens, at West 216th Street and the Harlem River and West 135th Street and Gansevoort Street in Manhattan, at Apollo Street and Newtown Creek and at Bay Street and Jamaica Bay in Brooklyn, and at Hunts Point in the Bronx.

Meanwhile, in addition to replacements, upgrades, and extensions of older plants, the department of public works built new sewage treatment facilities at Hunts Point in the Bronx, in Rockaway in Queens, at Owl's Head in Brooklyn, and at Port Richmond in Staten Island.

Water Supply

With four of its five boroughs situated on islands surrounded by saltwater, New York City relies on aqueducts and tunnels to bring freshwater from upstate. As the city grew and consumed more water, it suffered periodic droughts during periods of light rainfall.

In 1949, New York experienced a severe drought that forced the city government to consider constructing new ways to increase the city's water supply. Levels in the Croton and Catskill reservoirs declined to just 34.8 percent of their total capacity, compared to a normal level of nearly 80 percent. Assuming normal usage levels, the city's water supply would have held out for only 64 days before the system ran dry.

The city adopted emergency measures to save water. On December 16, the city mandated its first "dry Friday."[13] Men were asked to forgo shaving, families were discouraged from bathing, dishes were to be washed one at a time, and laundry was supposed to be washed only in large loads. New Yorkers were encouraged to drink as little water as possible, leading some people to comment that this was one more reason to be grateful that Prohibition had ended. The next week and the following month, the day was changed to "thirsty Thursday." These measures reduced the city's water usage by 300 million gallons a day. But spot surveys suggested that voluntary compliance was decreasing each time the special measures were demanded. As the drought continued, some freshly shaven men claimed that they had used electric razors or recycled the water they had used to wash their faces.

The situation demanded a more dramatic response. Chiding the city for allowing the emergency to arise, the Interstate Commission on the Delaware River Basin grudgingly granted New York permission to build an emergency plant on the eastern bank of the Hudson River, 60 miles from the city line, from which it could pump river water into its reservoirs. However, the commission's permission was conditional: Hudson River water could be pumped only in the event of a dire emergency. The plant would have to be dismantled by 1957, when, it was assumed, New York City would have completed a new system linking its water supply to the Delaware River.

The Neversink Reservoir (begun in 1941) and the Pepacton Reservoir (begun in 1947) were completed in 1954. In 1957, before it had ever been turned on, the Chelsea pumping station was dismantled.

Yesterday's Future

In 1961, a growing and confident New York City adopted a new zoning ordinance. It emphasized the creation of open space, included parking requirements, and coordinated use and bulk regulations. Introducing incentive zoning, it offered a bonus of extra floor space for high-rise office and apartment buildings that incorporated public plazas into their projects.

As with many public policies, the new zoning ordinance had intended—and unintended—consequences. New high-rise office buildings sprung up in the city's business districts. On the edges of the city, residential densities were reduced. But the emphasis on open space sometimes resulted in towering buildings that overwhelmed and were isolated from their surroundings.

Reflecting the best thinking of its time, the zoning ordinance served a city that had grown from about 5 million residents in 1900 to more than 8 million in 1960 and had emerged as the world's financial, cultural, and communications capital, as well as the headquarters of the United Nations.

While water supply remains a continuing challenge, New York used the prosperity of the postwar era to build a physical infrastructure and strengthen the social fabric that would allow the city to survive the economic ups and downs and the wrenching social changes of the decades to come.

Many city government officials—including its veteran water supply engineers and Robert Moses—shuddered at pouring polluted, chlorine-treated Hudson River water into their system, which was filled with fresh mountain water. (One water engineer said, "We're like a restaurant owner who has refused to serve his customers cheap wines but finds he has to bend with the times."[14] For his part, Moses said he was opposed to the Hudson River water for the same reason that he didn't like eggs that were only "fairly good.") But there were others who saw the use of Hudson River water not only as fiscally prudent but also as a crucial precaution during a dangerous time. Thus, the Citizens' Budget Commission argued that, instead of just pumping a hundred million gallons a day on an emergency basis, the city should dam the Hudson and make it a primary permanent supply source. Indeed, *The New York Times* pointed out that the Hudson River could be tapped for water if a new war once again shut down construction on the Delaware River system. Moreover, in the event of a nuclear attack, the Hudson River's water supply would be easier to decontaminate than smaller, contained waterbodies flowing through longer, enclosed channels.

In April 1950 work on the emergency pumping station got underway. Work crews put up the building and installed the pumping equipment on the river bank, while other crews lowered sections of pipe into the channel and divers bolted the pipe sections together. But, by the time the work was completed in May 1951, the worst of the drought was over. The Rondout Reservoir, which had been begun in 1937, was just being filled for the first time.

Bowery Bay Interceptor Sewer and Treatment Plant Extension (above)

New sludge storage tank for the Bowery Bay Sewage treatment plant takes shape. Capacity of the plant will be tripled when current work is completed. Construction started in 1955 on extension of the treatment plant and the Bowery Bay Interceptor Sewer, which will remove pollution from a big stretch of the East River, from 20th Street, Astoria, to Newtown Creek. Mayors Annual Report, 1955. Courtesy Municipal Archive Collection.

Owls Head Sewage Treatment Plant (opposite page)

Aerial view. Bay Ridge, Brooklyn. 1950–1952. Courtesy Municipal Archive Collection.

Underpinning

In New York City, putting up new buildings is only half the challenge. Most of the city is already so densely built up that there's another problem—making sure that the buildings near the construction site don't collapse.

Many areas of the city have difficult conditions just below the streets. Many miles of utility lines are tightly packed underground. Meanwhile, alongside the new buildings are many venerable and vulnerable buildings of considerable historic, aesthetic, or functional significance. Even when vibrations from blasting or from pile-driving don't create risks, excavations next to or beneath existing buildings can cause their walls to bulge, break, tip, or twist. Construction dewatering—lowering the water table in the vicinity of an excavation by drawing water and, sometimes, entrained sediments, from the soil—can further destabilize adjacent edifices.

Underpinning is the art and science of keeping existing structures intact while new ones are being built. As the word suggests, the basic principle involves the insertion of new foundation piles below existing buildings—something easier said than done!

As Stanley Merjan of the Underpinning and Foundation Company explains: "Underpinning is producing a new foundation under an existing building, where, for one reason or another, the existing foundation does not suit the new conditions. If a building was on a foundation alongside a site that is being developed, you needed to go beneath the foundation to support the existing building."

Some of the earliest underpinning efforts took place in New York City as its contractors struggled to build the first subways and the first skyscrapers. Relatively simple timber braces were angled against existing walls or wedged against the lagging boards that lined excavation pits.

These efforts were not always successful. When new subways were built early in the century, there were building collapses in Manhattan and Brooklyn.

Another early type of underpinning involved "needle beams." These usually were steel "I" or "H" beams punched through an existing wall near its base and attached to its support columns. The ends of these beams, projecting at right angles from the wall, rested on piles that extended far enough below grade to provide support. Although needle beams were relatively easy to install, they took up a lot of space. This made them difficult to excavate or build around. Moreover, the beams had to be removed in order to complete the new structure and restore the old structure to its prior use.

But with a continuing urgent demand for underpinning improvements, significant innovations were made during the 1910s as the second wave of subway construction got underway. The old-fashioned transverse needle beams were replaced with parallel beams that ran on either side of an existing wall and supported short cross beams that were pushed through the wall. These parallel beams in turn rested on grillage piles—footings composed of layers of relatively short, crisscrossed I beams—that extended far enough below grade to bear the wall's weight. These systems took up much less space (which made working around them easier) and, in many cases, the beams could remain in place permanently.

Centre Street, Manhattan (opposite page) BRT Centre Street, October 1909. Courtesy New York Transit Museum.

Some of the most complex underpinning challenges involved keeping one transit system—an el or an existing subway—in place while a new subway tunnel was bored beneath it. The contractors who built the "Dual Contract" and "Independent" subways from the 1910s through the 1940s faced this problem often. They held up the Eighth Avenue el while digging the IND beneath it, the Sixth Avenue el while digging the Sixth Avenue subway, or bored the IND and Sixth Avenue lines through midtown and downtown under existing IRT tubes. Because they were not digging through virgin territory (unlike the builders of the upstate aqueduct system, who generally tunneled through relatively undeveloped property), underpinning existing structures made subway construction at least 16 percent more expensive than it would otherwise have been.[1]

Sixth Avenue Subway (opposite page)
Construction was carried on under the street while the traffic moved overhead. The work was performed by Slattery Contracting. Courtesy Skanska USA Civil Northeast.

Sixth Avenue Subway (above)
The work was performed by Slattery Contracting. Courtesy Skanska USA Civil Northeast.

Timeline: 1963–1979

1963	1964	1965	1966	1967	1968	1969	1970
Pennsylvania Station demolished	Verrazano-Narrows Bridge	South Branch Interceptor Sewer	Rikers Island Bridge	Delaware Aqueduct System	Triborough Bridge and Tunnel Authority is merged into MTA	West Branch Interceptor Sewer	Cross Bay Bridge
Van Wyck Expressway	World's Fair	Con Ed Ravenswood Plant	Eastchester Bridge completed				Construction of Water Tunnel No. 3 begins
Cross-Bronx Expressway	Brooklyn-Queens Expressway	Cannonsville Reservoir	Mayor John V. Lindsay				
Hawtree Basin Bridge Completed	Dr. Martin Luther King Jr. Expressway		Richmond Tunnel				
Alexander Hamilton Bridge	Staten Island Expressway						
Clearview Expressway	Harlem River Drive						
New NYC Highways Department takes over from Borough Presidents							
Department of Public Works takes over sewer construction role from Borough Presidents							

1971	1972	1973	1974	1975	1976	1977	1978	1979
West Street Interceptor Sewer	Long Island Expressway	World Trade Center		Fiscal crisis	Battery Park City landfill	Eltingville Interceptor Sewer	Department of Environmental Protection is formed, subsuming Board of Water Supply as well as sewers	Northern Boulevard Bridge under construction
Silver Lake Storage Tanks	Korean War Veterans Parkway	Passenger Ship Terminal opens at Pier 88-90		Richmond Terrace Interceptor Sewer	West Shore Expressway			Newtown Creek Sewage Treatment Plant
	West Side Interceptor Sewer	Remaining Bronx 3rd Avenue, Webster Avenue El between 149th Street and Gun Hill Road closed					Mayor Edward I. Koch	
		Brooklyn Union Gas terminal for liquified natural gas at Smoking Point, Staten Island					Red Hook Interceptor Sewer	
		Con Ed agrees to halt the use of coal for generating electric power; by 1974 fuel oil produces 85% of Con Ed's power						

"There is a lot of competition. New York City demands the best and gets the best. Because of the competition, the industry keeps getting better. If you don't improve in what you do, you're not going to stay in business."

Thomas Iovino, *CEO and Founder, Judlau Contracting*

1963–1979: Urban Crisis

New York City Hits Bottom

For its first 300 years, New York City kept growing, building, and setting its sights as high as its skyscrapers. In the face of poverty, corruption, world wars, recessions and depressions, and racial, ethic, and class conflicts, New Yorkers maintained their faith in the future. But, from the late 1960s through the 1970s, something changed in the city's spirit as well as its economy and demography. In addition to losing people, businesses, and jobs and suffering from a fiscal crisis, New York City was losing its sense of its special destiny as the greatest city in the nation and the entire world. The city's economic, social, and spiritual crises were reflected in its infrastructure. In a series of alarming incidents, several of its bridges, highways, and other public facilities were crumbling and collapsing. Meanwhile, maintenance and repair projects were delayed, and new construction projects were deferred.

In fact, the city's decline had been building since what seemed to be a high point for New York and the nation—the years immediately after World War II. Many of the policies that built the mass middle class and the postwar prosperity, including the federal highway system and subsidies for home ownership, encouraged families and companies to flee the cities for the suburbs. Because of the high costs of doing business in New York and other cities, factories relocated from the urban centers to the suburbs, rural areas, and other countries.

On December 15, 1973, a section of the West Side Highway collapsed under the weight of a tractor-trailer truck carrying 30 tons of asphalt. The truck and a passenger car both fell to the street below, injuring the drivers, both of whom recovered. The highway caved in because its steel supporting beams had rusted and weakened. The next day, the highway was indefinitely closed south of 18th Street.

On June 28, 1981, two 600-foot suspension cables on the Brooklyn Bridge snapped, curling downward and tearing a hole in the wooden pedestrian walkway. One cable hit a freelance photographer, Akira Aimi, who had been walking on the footpath. His skull was shattered, and he died a week later. Consisting of two-inch thick bundles of galvanized wires, the cables had been eaten through by rain, snow, and pigeon excrement and had not been properly inspected and maintained. Fortunately, the cables were not needed to support the bridge's roadway.

Civil Rights Rally (above)
Civil rights advocates march in the Harlem section of New York to protest recent racial violence in Alabama. March 16, 1965. Courtesy National Archive/ Newsmakers, Getty Images News/Getty Images.

Lincoln Center (opposite)
An aerial view of the newly built Lincoln Center in Upper West Side, Manhattan. The complex includes the New York State Theater, Philharmonic Hall, Guggenheim Bandshell, Metropolitan Opera House, Vivian Beaumont Theater, and Julliard School of Music. Industrial buildings and the Hudson River are behind the complex. The foundation for Lincoln Center was built by GCA member John Civetta & Sons. 1968. © Charles E. Rotkin/CORBIS.

Mezzanine Area of the Welfare Island Tunnel (below)
Survey photograph showing the general view of the work progress in the mezzanine area of the Welfare Island tunnel for the construction of the 63rd Street Tunnel. January 7, 1971. Courtesy New York Transit Museum.

Thus, the 1960 Census reported an almost unprecedented development. For the first time since the nation's founding, the population of New York City had declined by 100,000 since 1950. Meanwhile, between 1960 and 1975, more than 600,000 factory jobs left the city, and, by the beginning of the 1970s, the city was losing 100,000 jobs of all kinds—white-collar as well as blue-collar—every year. Meanwhile, many of the newcomers to New York City were unprepared for the post-industrial economy and required assistance from government programs, including public assistance and public hospitals.

With the tax base declining and the city's budget increasing by 16 percent a year by the late 1960s, the city was having trouble balancing its books. Increasingly, the city resorted to "creative accounting." Hundreds of millions of dollars of day-to-day operating expenses that should have been in the expense budget, which should have been paid for with

tax revenues, were placed in the capital budget, which was paid by borrowing. This fiscal legerdemain crowded out the funds for construction projects, which would otherwise have been provided by the capital budget, and ultimately cost the city the confidence of the financial community.

By 1975, New York City was on the verge of defaulting on its debts and declaring bankruptcy. Turning for help to the federal government, the city at first was turned down by President Ford, whose speech rejecting the request for a loan was immortalized in the *New York Daily News'* famous headline, "Ford to City: Drop Dead." Later, the city did receive $2.3 billion in loan guarantees in return for strict controls on its spending and accounting.

As *New York Times* reporter Clyde Haberman wrote, largely as a result of the fiscal crisis of the 1970s, "Nothing of consequence ever seemed to get built in those days."[1] Before the fiscal crisis, New York City had invested over a billion dollars a year in construction contracts. But, by 1978, the city awarded only about $300 million in construction contracts, most of which covered completing projects that had been started in earlier years.

These were hard times for the construction industry. As GCA President James Moriarty Jr., President of T. Moriarty and Son, explains: "The '70s were a tough time for our company. The workforce that we had kept at a steady level for many years dropped off for lack of work. It wasn't that we were looking to lay off. It was just that fewer jobs were available." Together with others from business and labor, the construction industry did its part to restore the city to fiscal health. "During the fiscal crisis in New York, I was active in the GCA," recalls Ted Civetta of John Civetta and Sons, Inc. "Public works suffered. We bought bonds to promote specific projects and state bonds—the 'Big MACs' (Municipal Assistance Corp.)—to get the city out of the crisis. We survived."

Queensboro Bridge (left)
Survey photograph with views of the Queensboro Bridge and showing the construction equipment and excavation in progress on Welfare Island for the construction of the 63rd Street Tunnel. April 3, 1970. Courtesy New York Transit Museum.

Verrazano-Narrows Bridge (below, left)
Construction of the Verrazano-Narrows Bridge. 1963. Courtesy Skanska Civil Northeast.

Verrazano-Narrows Bridge (below, right)
Opening day motorcade at the Verrazano-Narrows Bridge. Photographer unknown. November 21, 1964. Courtesy MTA Bridges and Tunnels Special Archive.

To be sure, many of the transforming movements of the 1960s and 1970s exerted a positive impact on the nation, on New York City, and on the construction industry. The civil rights movement and the women's movement not only opened new opportunities for millions but also eventually opened up the construction industry to African Americans, Latinos, and women. The environmental movement not only made the nation and New York more conscious of the need for clean air, clean water, and sustainable uses of land and energy but also created a new demand for more responsible construction practices and projects such as mass transit systems and sewage treatment plants. Wrenching as the changes and challenges of the 1960s and 1970s undoubtedly were, they also sewed the seeds for renewing and rebuilding New York City in the 1990s and the first decade of the twenty-first century.

Bridges and Highways

At the beginning of this period, the long-awaited link between Brooklyn and Staten Island—the Verrazano-Narrows Bridge—was completed in 1964. Connecting the still sparsely settled bedroom communities of the smallest borough with the rest of New York City, the Verrazano-Narrows Bridge contributed to Staten Island's growth. A breathtaking sight, its towers soar 650 feet above the water, and its double-deck roadway is suspended 226 feet high. For 17 years, it was the longest suspension bridge in the world.

Queensbridge Park Tunnel
(above)

Survey photograph showing a general view of the work progress within the Queensbridge Park tunnel for the construction of the 63rd Street Tunnel. January 7, 1971. Courtesy New York Transit Museum.

Lincoln Center
(above, right)

Slattery was part of a joint venture that built Lincoln Center for the Performing Arts. Courtesy Skanska USA Civil Northeast.

World's Fair Train Car
(opposite page)

Photograph of an R36 World's Fair car in Queensborough Plaza Station. 1964. Courtesy New York Transit Museum.

The only other new bridge built during this era connected Queens with the former landfill—and current penitentiary—on Rikers Island. The Rikers Island Bridge was begun in 1964 and completed in 1966. Two other major bridges built during the 1960s replaced existing crossings: the Eastchester Bridge in the Bronx and the Cross-Bay Bridge in Queens.

Meanwhile, a variety of expressway projects were proposed, including Robert Moses' plans for three cross-Manhattan expressways and an expressway through Bushwick in Brooklyn. But these projects were halted by opponents from these communities. Meanwhile, the Van Wyck Expressway in Queens was finally completed in 1963.

The 1964 World's Fair

In another example of how the era was unfavorable to major projects, the 1964 World's Fair was not a financial success, nor did its chairman, the ubiquitous Robert Moses, succeed in using the exposition to produce a new generation of infrastructure products. Using the Flushing Meadows Park in Queens, the fair opened on April 22,

1964, and ran for two six-month seasons concluding on October 17, 1965. While more than 51 million people attended the fair, this turnout was much less than the anticipated 70 million.

The Environmental Movement and Its Consequences for Construction

During the 1960s and 1970s, environmental activism called attention to the deteriorating air quality, nationally and in New York City. Although New York's air quality was not as bad as in Los Angeles or many industrial cities, the city Department of Air Pollution Control's laboratory measured unacceptable concentrations of pollutants.

As the environmental movement gained strength, New York City's Bureau of Air Resources took several dramatic actions against pollution. First, the bureau banned the use of soft coal for heating residential and office buildings. Then, it closed tens of thousands of apartment house incinerators (which the city had formerly required in every building with 12 or more units). Finally, in compliance with regulations enforcing the new Federal Clean Air Act, the city closed 8 of its 11 municipal incinerators.

First Earth Day
(left)

Earth Day was heeded by many who turned out in appeal for a regeneration of a polluted environment. The celebration, in which millions across the nation participated, was an exuberant rite of spring. On Fifth Avenue, one group of marchers carried a tree. New York, NY. April 22, 1970. © Bettmann/CORBIS.

The first Earth Day—April 22, 1970—increased awareness of the benefits of "recycling" (a term that originated in 1963). Neighborhood recycling centers, some of which would stay in operation for more than a decade, were complemented by the city's first experiments since World War II with collecting different kinds of wastes that could be reused—"source-separated recyclables," as it was referred to.

As with all Americans, New Yorkers became more conscious of the need to conserve energy after the oil embargo imposed by the Organization of Petroleum Exporting Countries on October 17, 1973, resulted in long lines at gas stations, shortages of heating oil, and layoffs at many companies. Even more dramatically, the dangers of the growing demand for energy were brought home—literally—on the evening of July 13, 1977. When a lightning strike at 8:37 p.m. knocked out a substation of the giant electrical utility company Consolidated Edison, the incident set off a chain of events that plunged New York City into a power blackout. In the chaos that followed, 3,776 people were jailed (the largest mass arrest in city history), 1,616 stores were looted, and at least 1,037 buildings were torched. The total damages were estimated at $300 million.[2] For years to come, the power failure convinced millions in New York and throughout the nation that energy must be conserved, the city's social fabric was fraying, and many inner city neighborhoods were in crisis.

Hudson had been debated as a less expensive alternative to building the Cannonsville Reservoir, the Board of Water Supply ultimately chose the reservoir course, which was built from 1955 through 1965. With its completion, the long awaited Delaware Aqueduct system was finally complete. The siphon under the Narrows to deliver water to Staten Island was replaced from 1964 through 1966, and giant storage tanks for Staten Island's water supply were built at Silver Lake from 1967 through 1971.

Sewage Treatment

Environmentalism also encouraged improved treatment of liquid and solid wastes, helping to spur several construction projects. After 14 years of construction, the Newtown Creek Sewage Treatment Plant in the Greenpoint section of Brooklyn was completed in 1979. The facility now handles sewage from Brooklyn, Queens, and Manhattan.

As the economy recovered, New York City decided to do something about the 170 million gallons a day of raw sewage flushed into the Hudson River from the west side of Manhattan, from its northern tip to 14th Street. During the 1920s and 1930s, city officials planned to pipe the sewage to Wards Island. After Department of Public Works engineers discovered that the cost of this tunneling would be too high, they investigated potential locations along the Hudson River. They first proposed a two-block site on the Hudson River at West 72nd Street. But, by 1965, a new site had been selected between 137th and 145th Streets, and work on this North River plant site began in 1972. One of the contractors involved in this project was GCA member Perini & Sons, Inc. Meanwhile, work on the West Side Interceptor Sewer, which would funnel the flow from West Side sewers to the plant, began in 1967 and was completed in 1972.

Water Supply

As often occurred during New York's history, the city thirsted for new water supply, and the means of delivering water were in need of repair and replacement. Although damming the

As early as 1954, the Bureau of Water Supply had decided to inspect Water Tunnel No. 1. But, as engineers tried to close the six-foot-diameter valve to shut off the water entering the tunnel, the brass turn shaft started to shudder and crackle. "They were afraid," a worker later told a reporter, "if they turned it any more the whole ... thing would break." With no way to turn off the water, there was no way to know whether or not there was a condition that might cause the tunnel to burst, or to repair a crack even if they had found it.

The only way to inspect and repair Water Tunnels No. 1 and 2 was to build a third tunnel that would extend 60 miles from the reservoir in Yonkers through the Bronx and Manhattan and into Brooklyn and Queens. This tunnel could be operational while the other water tunnels were tested. By the late 1960s, designs were prepared, and contracts were ready to build. Work began in January, 1970, but, in the depths of the fiscal crisis in 1975, Mayor Beame ordered a halt. Slowly and fitfully, the work resumed in 1979, but full-scale work would not begin again until the early 1980s as the economy improved and funding was possible.

North River Water Pollution Control Plant (above)
Foundation for North River Water Pollution Control Plant built by GCA member Perini Construction. Other GCA members involved in this project include Nicholas DiMenna & Sons, Inc; Poirier & McLane; Andrew Catapano Co., Inc.; NAB Construction Corp; Slattery Contracting Co.; Schiavone Construction, and Corbetta Construction Co.
Courtesy Perini Corporation.

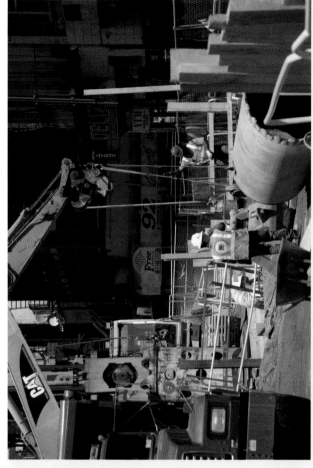

Second Avenue Subway (opposite page)
View of east side Second Avenue looking south showing subway construction on Route 132A Section 13. Photograph shows several workers in trench. GCA members Thomas Crimmins Contracting, Cayuga Construction Corp., and Slattery Contracting worked on this project. December 24, 1973.
Courtesy New York Transit Museum.

Second Avenue Subway (right)
Construction of the Second Avenue Subway from 91st to 96th Streets. 2008.
Courtesy S3 Constructors, a joint venture of Skanska, Schiavone and Shea.

Notes from Underground

Even more than in other eras, most of the technological innovations in building and rebuilding the city's infrastructure were difficult to see or appreciate because they occurred underground. In the most famous (and most difficult) excavation project of this period, was the foundation for the giant Twin Towers at the World Trade Center in Lower Manhattan. Other foundation innovations included the increased use of narrow-diameter "pin" and "micro" piles, which could be used in a wider variety of applications, at considerably less cost, than standard pile technology. By imparting high-frequency vibrations to steel piles, "sonic" piledrivers could force them into the ground with greater speed and efficiency than with traditional techniques. Moreover, the sonic piledrivers were virtually noiseless, eliminating what had been a major nuisance in many neighborhoods.

Demolishing Historic Landmarks

Meanwhile, above ground and seemingly symbolizing the era's iconoclasm, several of New York City's best known and most visible landmarks were being demolished and rebuilt or relocated.

Completed in 1910, the original Pennsylvania Station was a magnificent monumental gateway to New York City. Located on Eighth Avenue between 31st and 32nd Streets on Manhattan's west side, Penn Station was built partly with pink granite and boasted an imposing colonnade of columns in the classical Greek and Roman styles. Its granite and marble walls were modeled after the famous Roman Baths of Caracalla. Beginning in 1964, the original station's above-ground building was torn down and replaced by the Pennsylvania Plaza complex, including the new Madison Square Garden. "The sadness and resentments will be replaced in time when the populace experiences the conventions and the joys that the new facilities will provide," wrote an anonymous correspondent for the Bulletin of the General Contractors' Association in 1967, as the new building was taking shape. But the writer was wrong. New Yorkers mourned the passing of the Penn Station that they had known and loved, and the furor over the demolition of this landmark prompted the creation of the New York City Landmarks Preservation Commission and boosted the national movement for historic preservation.

An active coalition of citizens led by Jacqueline Kennedy Onassis and the Landmarks Preservation Commission thanks to the Landmarks Preservation Act, passed in April 1965, and upheld by the U.S. Supreme Court in 1978, made sure that New York's other landmark from the golden age of railroads—Grand Central Terminal—was saved from demolition in 1978.

Another historic site that was shuttered was the Washington Market, the venerable complex of wholesale establishments on Manhattan's lower west side. Served by rail and carfloat, the market provided the city with much of fresh fruit, vegetables, meat, and dairy products. After most of the merchants moved to a new city market at Hunts Point in the South Bronx, Washington Market was replaced by high-rise apartments and a new community college.

TWA Terminal at JFK Airport (left)
The futuristic look of the TWA terminal at JFK Airport. The TWA terminal is deserted by passengers who refuse to take a TWA flight due to bomb threats. Photographer JP Laffont. © JP Laffont/Sygma/CORBIS.

TWA Terminal at Kennedy Airport (opposite page)
The TWA terminal at Kennedy Airport. New York, New York, 1967. Photographer W. Wayne Lockwood. © W. Wayne Lockwood, M.D./CORBIS.

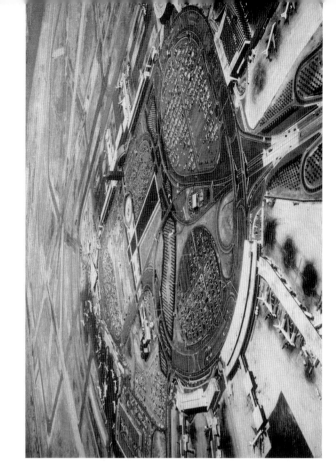

Idlewild Airport
Idlewild Airport. Aerial view of new runways. October 1961.
Courtesy Municipal Archive Collection.

Mass Transit

The environmental movement supported mass transit, and an ambitious new transportation project began but was not completed during this era. On October 25, 1969, the city received a $99 million grant from New York State to build the Second Avenue Subway and its associated 63rd Street Tunnel, both on the east side of Manhattan.[4] However, a New York State transportation bond issue that, among other objectives, was intended to continue funding the Second Avenue Subway was defeated by the voters in November 1971, carrying the city but losing elsewhere. In June 1972, a $25 million grant from the federal Urban Mass Transit Administration kept the project alive. But, in November 1974, in the midst of the fiscal crisis, Mayor Abraham Beame halted work on the project after only a mile-and-a-half of the north-south tunnel had been bored. This perpendicular 63rd Street Tunnel, connecting the new line to Queens under the East River, was eventually completed in 1989.

The World Trade Center

In the major construction project of this or any other era, work began in 1966 on the World Trade Center—seven buildings in lower Manhattan, including the iconic 110-story Twin Towers. So ambitious was the project that the place where its excavation spoils were placed turned waters that had once been used for piers into a landfill on which Battery Park City—a planned community on the tip of Manhattan Island—would be built.

World Trade Center Under Construction (this page and opposite page)

The World Trade Center towers, the second largest buildings in the United States, under construction in New York City. GCA member Karl Koch Erecting (now Skanska Koch) built the structural steel frame for the World Trade Center, February 1971. Top right photo © Charles E. Rothkin/CORBIS. Other photos: Courtesy Port Authority of New York and New Jersey.

GCA contractors played important parts in building the World Trade Center. As John Donohoe, who now serves as chairman of Moretrench, recalls; "I was the first superintendent that Moretrench put in charge of the original dewatering of the World Trade Center. This had to be done carefully. We installed a dewatering system to dewater the south tube of the PATH system into the excavation that was being made. If the water was drawn down too rapidly, the soft, compressible materials that the tunnel sat on would settle and cause the tunnel to move. If the water was not drawn out rapidly enough, the tunnel would float up."[5]

When the World Trade Center opened in 1973, it was a sign of hope in the midst of the city's crises. Three decades later, in the aftermath of the city's greatest tragedy, it would once again symbolize the city's resilience and renewal.

Concrete

If New York, or any modern city, could be said to be built of any one material, it would be concrete. Its heavy infrastructure—its roads, sewers, subways, sewage treatment plants, and such—consists largely of concrete. The Hoover Dam on the Colorado River—more than 700 feet high, more than 1,200 feet wide at the top, more than 660 feet thick at the bottom, and weighing about 8 million tons—is the largest concrete structure in the world. As David Owen points out, New York contractors pour a new Hoover Dam's worth of concrete on the city—most of it beneath street level—every 18 months. [1]

Though it also played an important part in construction during the Roman Empire, the secret of making concrete was lost for millennia. In 1756, the British engineer John Smeaton pioneered the use of hydraulic lime in concrete, mixing it with pebbles and powdered brick as aggregate . Soon builders were heating limestone and clay to drive out the water and then pulverizing it to form a powder (cement) that can be mixed with sand, crushed stone, and water to form a strong, highly durable material (concrete) that can be molded into almost any shape and bound to almost any material. Because of its long-lasting compressive strength, its versatility (it was, if anything, even stronger under water than on land), and—perhaps most importantly—its low cost, concrete rapidly assumed an increasingly important role in construction. Before the end of the nineteenth century builders had discovered that concrete's tensile strength could be dramatically improved by embedding steel rods (rebar) in the wet concrete before it set. By the beginning of the twentieth century such "reinforced concrete" was widely used in building New York's infrastructure.

Ransome Concrete Mixer (above)

General Ransome, who invented this concrete mixer, helped manage construction of the Panama Canal. Ransomes were among the mixers used to build the Catskill Aqueduct system. October 11, 1890. Courtesy Scientific American.

> "Every time I hear about a water main
> break, it seems it was built in 1897 and
> is still in use over 100 years later. That
> doesn't make any sense. Infrastructure
> is critical to our economic well being
> and a continuous investment is needed,
> regardless of economic times."
>
> Joseph Ferrara, *President / CEO,*
> *Ferrara Brothers Materials Building Corp.*

Soon there were changes in how concrete was mixed, transported, and used. Concrete is unusually versatile. It can be molded into a weight-bearing structure of almost any shape. It can be joined to a wide variety of materials. And its physical characteristics can be modified to meet a range of site-specific conditions. It is so adaptable because the ratio of its basic components (cement, sand, aggregate, and water) can be adjusted every time it is mixed, and the wet mix can be poured (or shoveled or sprayed) into molds (forms) of any shape. But wet concrete begins to harden immediately and therefore has to be placed as soon as it is mixed. This means that the mixing has to take place either at or very near the construction site. At the beginning of the twentieth century, bags of cement were mixed with sand, aggregate, and water in troughs and stirred by hand with shovels until the cement and sand were evenly moistened and blended and this mortar had evenly coated the gravel aggregate. This was hard, heavy work.

It was also slow. And despite conscientious efforts to ensure that each small hand-mixed batch was the same as the last, it was virtually impossible to maintain a uniform consistency.

Many inventors therefore tried to harness the power of steam to make concrete mixing faster and more efficient. The Panama Canal (completed in 1914), which required the application of vast quantities of concrete, was a major spur to the development of improved mixing technology. So was New York's Catskill water supply system, with its enormous dams and tunnels. This aqueduct system provided an unequaled field lab for simultaneously testing many of the latest advances in concrete mixers. This equipment was used to create large-scale mixing plants that could produce all or most of the concrete for a specific major facility (such as the Ashokan Dam) and be erected very near the construction site.

Concrete also required forms to hold the wet mixture in the desired position until it had set. The earliest forms were made of timber. While wooden forms offered flexibility—they could be built on-site to meet almost any specific condition—and the wood itself was relatively inexpensive, they were nonetheless quite costly because it was difficult to reuse the lumber more than a few times and because building them and taking them apart each time was a time intensive endeavor. Given the inevitable variations in the lumber and in its construction, it was also difficult to ensure the consistency of the finished surface.

Steel forms, on the other hand, were virtually indestructible. Therefore, they could be reused indefinitely, they could be fashioned in portable sections that could be readily transported and quickly assembled and disassembled, and they offered complete uniformity in the finished concrete surface while still being adaptable enough to be installed in almost any situation that might be encountered on a specific project. To avoid their relatively high purchase cost, they could even be leased. Firms such as Blaw-Knox [a GCA member], which was one of the first to design portable, reusable steel forms, quickly became the predominant suppliers of the form equipment used by New York contractors to build the city's basic infrastructure.

Despite making the mixing and application of concrete more efficient, the essential raw material—cement—was always packaged, transported, and stored in bags. This made handling concrete inherently labor-intensive. Individual bags had to be loaded by hand into boxcars, unloaded by hand, taken to an interim storage location, and finally opened and emptied into the concrete mixer. The irreducibility of this fundamental unit—the bag that contained one cubic foot of concrete—was reflected in the designations for each size and type of mixer: they were either one bag, two bag, or three bag machines.

East River Drive (above)
This relatively elaborate chute system shows another way to get wet concrete from the bucket that conveyed it from the mixing plant into the form. East River Drive Contract 7, 54th to 64th Streets. October 10, 1939. © New York City Department of Parks

East River Drive (above)
Pouring concrete for the Northend foundation overpass on the East River Drive. September 6, 1940.

FDR Drive Rehabilitation (below)
John Picone Inc. built and operated the first large water floating concrete plant in New York history. By using the floating concrete plant, Picone was able to make continuous pours of up to 2,500 cubic yards, and complete a highly complex project on time. Courtesy John Picone Inc.

Sand and gravel aggregate, on the other hand, had always been shipped, handled, and stored "in bulk." Rather than individual bags shipped in boxcars, sand and gravel were dumped loose into barges or into open-topped gondola railcars that were covered with a tarp to keep out moisture. The material was then tipped into a storage bin that was designed to allow its contents to flow out the bottom, by gravity, in regulated amounts. The first bulk cement plant in New York City may have been the Cranford Company's, built (so that the sand and gravel could be shipped by barge) on the shores of the Gowanus Creek in Brooklyn in 1930. (The Cranford Company was a GCA member.)

The switch to bulk cement—whose cost-effectiveness made the old bagged form obsolete almost overnight—led to the development of "ready-mix." This required not only bulk cement but also equipment that could mix and transport the material at the same time so that it could be used before it set.

Ready-mix technology depended on the development of truck motors that were strong enough to carry and power heavy rotary drums that could mix the concrete, keep it from setting until it was delivered, and then scrape it out of the mixer so that it could be chuted into forms. Various equipment makers experimented with a variety of truck-mounted rotary agitators until the current, universally recognizable concrete truck had evolved.

Even the modern concrete truck did not release the concrete industry from the need to site its batching plants in relatively close proximity to their center of demand. A load of concrete, even under constant agitation, becomes unusable after about 90 minutes. New York's concrete batch plants, therefore, generally remain on waterfront locations (so that they can continue to receive their heavy raw materials by barge) that are not far removed from the city's center.

A specialized concrete application is road paving. Automated equipment for paving roads with concrete also evolved over the course of the last century, although asphalt would become the primary paving material. "Concrete doesn't rot, and it doesn't burn," said Joseph A. Ferrara of Ferrara Brothers Materials Construction Corp. "If you produce and install concrete properly, you'll never have to replace it in your lifetime."

"With chemicals, we can make concrete as strong as steel," Ferrara continued. "We can take the production of concrete to a point where the engineers who design buildings have complete confidence that it can withstand anything. It is possible to produce concrete that can withstand pressure of 12,000 or 13,000 pounds per square inch."

Timeline: 1980–2009

1980	1981	1982	1983	1984	1985	1986	1987	1988	1989	1990	1991	1992	1993	1994

1980

DEP begins replacing 6-inch water mains

Red Hook Marine Terminal

1982

Metro-North is created by MTA to take over remnants of New York Central and New Haven Railroad suburban lines

1985

Harlem River Rail Yards

1986

North River Sewage Treatment Plant Opens

1987

Greenpoint Avenue Bridge

Red Hook Sewage Treatment Plant

1988

Ocean Dumping Ban Act

1989

63rd Street East River subway tunnel

1990

Nassau Expressway

Mayor David N. Dinkins

1993

Mole introduced into Water Tunnel No. 3

1994

Mayor Rudolph W. Giuliani

1995	1996	1997	1998	1999	2000	2001	2002	2003	2004	2005	2006	2007	2008	2009
Brooklyn Navy Yard Cogeneration Plant	Rikers Island Compost Facility	Ninth Street Bridge	Sunrise Highway		Howland Hook Marine Terminal	World Trade Center attacked	Mayor Michael R. Bloomberg	AirTrain	Shore Parkway Bridge	Third Avenue Bridge	Whitestone Expressway Bridge Rehabilitation	Second Avenue Subway restarts	Hugo Neu Recycling Facility	
			Oak Point Link			East Side Access Project begins	WTC cleanup ends			NYPA 500 megawatt plant	Kill Van Kull dredging underway	First Croton Funded Playground completed	Extension of #7 Line Subway to Javits Convention Center begins	
			IND Queens Blvd Line East River Tunnel							Croton Water Filtration Plant begins	Howland Hock Freight Rail Line completed			
			Water Tunnel No. 3, Stage 1, into service in August							First of more than 70 Bronx Parks reconstruction projects related to Croton Water Filtration Plant begins	Flushing Water Quality Facility			
											Staten Island Transfer Station			

1980–2009: Revival

Jacob Javits Convention Center (above)
Construction of the steel frame for the Jacob Javits Convention Center by GCA member Karl Koch Erecting. Built in 1986 and named after New York State Senator Jacob Javits who died that year.
Courtesy Skanska Koch.

East Side Access (opposite page)
Joint venture of GCA members Judlau Contracting and Dragados USA, Inc. Assembly of state of the art double shield tunnel boring machines in the bedrock below Manhattan's east side.

AirTrain (below)
Balanced cantilever guideway erection at JFK Central Terminal area. Courtesy Skanska Koch.

During the 1980s, 1990s, and the first decade of the twenty-first century, New York City once again displayed the vitality, ingenuity, and resilience that made it the world's leading metropolis.

Welcoming new immigrants, generating new businesses, and addressing the seemingly intractable challenges of public safety, environmental management, and fiscal deficits, New York City starred renewing itself again. As the city once more built and rehabilitated schools, subways, sewage treatment plants, housing, hospitals, and highways, New York refuted the naysayers who continued to believe that the metropolis was in an inevitable process of decline. When the worst tragedy in the nation's history struck New York on a clear September morning, the city set an example for the nation by uniting and rebuilding.

"If Less Is Better"

By the late 1970s, many academics, city planners, and public officials thought New York City's best days were behind it. According to the city's Housing and Development Administrator Roger Starr in 1976, it was time to embark upon a policy of "planned shrinkage."[1] In that spirit of accepting, even welcoming, decline the City Planning Commission later called for phasing out up to 5,000 beds at the municipal hospitals, putting a moratorium on new construction at the City University, and delaying the development of a new subway line in Queens.[2]

As a leading urban scholar, George Sternlieb, put it in 1976: "It takes a man who's been shot in the head a while to realize he's dead. New York may not realize it, but, if you look at the numbers, it's clear that New York is dead."[3] Sternlieb predicted that, in 20 years, New York City would have 15 to 20 percent fewer people.

If New York City was declining or even dying, then why build new public facilities? The prevailing pessimism made itself heard in dozens of controversies about construction projects. One of these debates was about whether to build new schools. Early in 1979, the City Planning Commission urged Mayor Koch to "reject all major new capital-construction projects," "deny funds requested for building public schools," and "close 200 public schools." In justifying this recommendation, the commission's chairman, Robert F. Wagner III (the son of the former mayor and grandson of a senator), declared that the city's "population is shrinking, its economy has contracted." Within about a decade, he predicted, public school enrollment would decline by 25 percent in the lower grades and by 40 percent in the high schools. In an editorial on April 15, 1979, *The New York Times* endorsed the commission's recommendation, declaring, "If less is better, the New York City of 1990 will be better."[4]

"Now we are working on the East Side Access Tunnel, which will connect the Long Island Railroad's Main and Port Washington lines in Queens to a new terminal underneath Grand Central Terminal in Manhattan. This new subway tunnel will go under the East River from Queens into the East Side of Manhattan."

Thomas Iovino, *CEO and Founder, Judlau Contracting*

Franklin Avenue Station (right)
Photograph documenting the conditions of the Franklin Avenue Station of the BMT Franklin Avenue Station of the New York City Transit Authority Maintenance of Way Division. Typewritten label on photograph reads: North end of Franklin Avenue Station. Burnt wooden platform at present closed to public. The rehabilitation of the Franklin Avenue Shuttle earned GCA member Judlau Contracting a "2000 Award of Merit: Transit Project" from New York Construction Magazine. March 26, 1981.
Courtesy New York Transit Museum.

Daily News "Ford to City: Drop Dead"
(above)
Courtesy Dailynewspix.com

Traveling on Foot
(opposite page)
April 1980: Thousands of commuters pour across Brooklyn Bridge on their way to work on the tenth day of the New York City transport strike.
© Keystone, Hulton Archive/Getty Images

Defying Decline

By the mid-1990s, New York City was surviving, thriving, and proving the experts wrong.

The enrollment in New York City's public schools did decline by 3.6 percent in the fall of 1979. But the decline in 1980 was just half as much, and, by 1982, the falling-off had halved again. Beginning in 1983, the city's school enrollment began to move upward, and, by 1990, it had begun to increase dramatically. Enrollments soon exceeded their 1979 levels, and the public schools soon faced problems of overcrowding, not excess capacity. Heeding the recommendations of city planners a decade earlier, the city had closed about 100 schools. But, by the early 1990s, about 90,000 schoolchildren were being taught in locker rooms, hallways, closets, trailers, and teachers' cafeterias.[5]

The city's schools were crowded because the city's population was growing. Yes, about a million people had moved out of New York City during the 1970s. But about a million new immigrants from 188 countries had arrived to replace them. And the city's overall birthrate had increased by 24 percent—twice the national average. As in earlier decades, the overwhelming majority of these immigrants had come to America—and to its gateway of opportunity, New York City—in order to work hard, start new businesses, and build better lives for themselves and their children. Neighborhoods that had been emptying out were being reborn as these newcomers moved in, mostly bringing with them valuable skills, entrepreneurial energy, and cohesive family structures. In 1988, the State Legislature established the School Construction Authority (SCA) to build new public schools and manage the design, construction and renovation of capital projects in New York City's more than 1,200 public school buildings, half of which were constructed prior to 1949.

Responding to the population growth, in 1985, Mayor Edward Koch announced a Ten Year Plan for Housing, which included a "five-year $4.4 billion program to build or rehabilitate around 100,000 housing units for middle class, working poor and low-income families and individuals." Three years later, the city government increased its investment to $5.1 billion and raised the number of publicly assisted units to 225,000.[6]

Vol. 57. No.109
····
Final

DAILY NEWS
NEW YORK'S PICTURE NEWSPAPER

New York, N.Y. 10017, Thursday, October 30, 1975

15¢

FORD TO CITY: DROP DEAD
Vows He'll Veto Any Bail-Out

Abe, Carey Rip Stand

Stocks Skid, Down Down 12

Three pages of stories begin on page 2, full text of Ford's speech on page 36

New Yankee Stadium
(above)

GCA members have a long history in building and rehabilitating Yankee Stadium. First built by GCA member White Construction, the stadium was rehabilitated in the 1970s by Karl Koch Erecting among other contractors. Renamed Skanska Koch, they are helping to build the new stadium for the New York Yankees.

Courtesy Skanska Koch.

and doing business. As often happens in New York City's history, there were clashes between the newcomers and the old timers. But, as in the past, these conflicts were signs of ferment as well as friction—and, feisty as ever, New York and its neighborhoods kept changing and growing.

By most indicators, New York City was on the upswing. Along with the influx of population came an increase in jobs. During the 1980s, there was a net growth of 370,000 jobs. Tax revenues tripled—from $879 million to $2.56 billion.[7] In an especially welcome trend, what had been a steady rise in crime finally began to flatten. Although the numbers of murders and auto thefts continued to climb each year through the 1980s, the numbers of robberies and assaults leveled off. Meanwhile, the number of rapes fell by 16 percent, burglaries by 42 percent, and thefts by 57 percent.

From 1980 through the stock market crash in October 1987, the city's real estate values soared. Soon after the crash, real estate values started increasing again. New owners bought the city's stock of "in rem" housing—buildings seized for tax delinquency, most of which had been abandoned and many of which were at risk for arson. And the city embarked on a $500-million program to build new subsidized housing for low- and middle-income people.

After 1990, the city's crime rates dramatically declined in every category. The number of murders declined by half between 1990 and 1995 and halved again by 1998. By 2005, there were 70 percent fewer murders than there had been in 1980, burglaries were down by 89 percent, auto thefts by 82 percent, other thefts by 81 percent, robberies by 76 percent, assaults by 60 percent, and rapes by 56 percent.[8] In 2006, with another 7.2 percent overall drop in the seven major crime categories, New York remained the safest of the country's 25 largest cities.[9] Public policies contributed to these improvements in public safety. In 1991, Mayor David Dinkins initiated the Safe Streets program which included hiring thousands of new police officers. Taking office in 1993, Mayor Rudolph Giuliani pursued the "no broken windows" philosophy of law enforcement, cracking down on seemingly minor offenses in order to discourage lawbreaking of all kinds. Under the Bloomberg Administration, the Real Time Crime Center provided the first centralized technology center for the Police Department, with instant and comprehensive information about crime patterns.

In the last decade of the twentieth century, New York City was once again becoming a more attractive place to live, work, do business, and raise families. More people wanted to live in the city, and the municipal government responded with new initiatives to promote affordable housing. Initiated in 2003 by Mayor Bloomberg and expanded into a $7.5 billion effort in 2006, the New Housing Marketplace Plan seeks to create affordable housing for 500,000 New Yorkers.

Around the country, many other old urban centers—for instance, Cleveland, Detroit, and St. Louis—continued to lose population, businesses, jobs, and tax revenues. Meanwhile, as in the past, New York City became a magnet not only for immigrants but also for energetic young people (including aspiring artists, actors, writers, and inventors) from throughout the country. In yet another population influx, young professionals from the suburbs—soon to be called "gentrifiers," or, more crudely, "yuppies"—also moved into the city. Together, these immigrants, professionals, and creative people of all kinds combined to preserve and revive one formerly industrial neighborhood after another. From SoHo and Tribeca and the East Village in Manhattan, to Williamsburg and DUMBO and Red Hook and Bushwick in Brooklyn, Long Island City in Queens, and Port Morris in the South Bronx, neighborhoods that had lost many of their original uses remained or were revived as places for living, working,

Construction Rebounds

Throughout New York's history, construction has been spurred—and been spurred by—the city's growth. As New York's population increased, its economy recovered, and its social conditions improved, new construction was commissioned and contributed to the city's revival.

New York City renewed its commitment to building and rebuilding its infrastructure. The capital budget (which consists mostly of funds for construction and repairs) for Fiscal Year 1981 was double the $400 million budgeted the year before. In 1982, the capital budget doubled once again.[10] Between the Fiscal Years 1984 and 1991, capital spending quadrupled to well over $5 billion a year. In 1990, facing an economic downturn, Mayor Dinkins reduced capital spending, but it still remained well over $4 billion a year.[11]

During other economic downturns in the years ahead the city would make other cuts in capital spending plans, even for such crucial items as bridge maintenance. But the overall direction of spending was upward. For instance, the four-year plan released in 2000, with $25 billion in capital spending between Fiscal Years 2001 and 2004, provides for the greatest investments in construction in the city's history.[12]

To be sure, important construction projects were still being delayed by spending constraints and other problems. Years of neglect had produced enormous needs for construction and rehabilitation that exceeded even the increasing allocations of capital spending. Moreover, the city's plans were restricted by a debt limit imposed by the New York State Constitution, which affected all city capital spending except the water and sewer work conducted by the Department of Environmental Protection.

In spite of these limitations, ambitious and long overdue projects began in mass transit, bridges and highways, sanitation and sewage, water supply, rail freight, and ports. And, when terrorists attacked, New York City responded to the challenge—and rebounded yet again.

Construction of Howland Hook Container Terminal (above)
1980. Courtesy Spearin, Preston & Burrows.

Jacob Javits Convention Center (below)
Structural steel frame for the Jacob Javits Convention Center was built by GCA member Karl Koch Erecting, 1983. Courtesy Skanska Koch.

Mass Transit

As the 1980s began, New York City's subway system was in urgent need of maintenance and modernization. The trains, the tracks, and the stations were breaking down, filth and graffiti were everywhere, and, for all their tolerance of discomfort and disorder, New Yorkers were complaining that the system was becoming unsafe, unreliable, and unpleasant.

Taking office in 1979 as the head of the Metropolitan Transportation Authority, the builder and developer Richard Ravitch began an ambitious capital improvement program. Tracks, switches, and signals were repaired and replaced. The major train yards in Brooklyn and the Bronx were modernized. Subway stations were renovated. And, by the end of the decade, a new fleet of stainless steel subway cars began running. This modernization program continued under Ravitch's successors, including David Gunn and Richard Kiley. Just as the city has become more liveable, the subways have become more ride-able.

For the first time, travelers can arrive at a New York City airport by rail. With construction beginning in 1998 and being completed in 2003, the eight mile, $1.9 billion AirTrain now runs between the Jamaica and Howard Beach Stations on the New York City subways and JFK Airport. Another project making it easier to travel by subway to a rail line is the 63rd Street subway tunnel under the East River that separates Manhattan and Queens. Originally, the tunnel was designed to connect to the long awaited Second Avenue Subway, as well as providing a lower level of track to

allow the Long Island Railroad (LIRR) to reach Grand Central Station. Mostly completed in 1976, the tunnel was finally put into service—for subways only—in October 1989.

In December 2006, the U.S. Department of Transportation made a commitment that the lower half of the 63rd Street Tunnel —the tracks originally intended for the LIRR—would also be used to allow LIRR riders to take their trains to Grand Central Station so that they could get to the East Side of Manhattan more easily. Although work had already begun to connect the LIRR's Sunnyside Yards in Queens to the East River Tunnel, the project was boosted by the $2.6 billion federal funding commitment.[13]

Almost a quarter century after construction was halted on the Second Avenue Subway, planning was resumed for the project in 2000. In December, 2006, the $3.8 billion project received a promise of federal funding. Construction on the subway began in 2007 and the long awaited dream of the Second Avenue Subway is moving closer to completion.[14]

Grand Central Terminal

While the demolition of the original Pennsylvania Station had outraged preservationists, the historic Grand Central Terminal was rehabilitated and preserved. In 1983, the station's operations became the responsibility of Metropolitan Transportation Authority's Metro-North Railroad. Afterwards, the railroad began to repair or replace much of the station, including a $4.5 million project to refurbish the leaking roof and skylights.

Cortlandt Street Station (below)
Jetgrout operations underway at the R/W subway line Cortlandt Street Station for the Dey Street Concourse Structural Box Project.
Courtesy of Skanska USA Civil Northeast.

**Utility Relocation
(opposite page)**
Utility Relocation in the Wall Street area, 2005.
Courtesy Judlau Contracting, Inc.

FDR Drive (right)
Rehabilitation of FDR Drive by Skanska USA Civil Northeast. February 14, 2003. Courtesy of Skanska USA Civil Northeast.

Second Avenue Subway (below)
During the building of a section of the Second Avenue Subway Line, Skanska USA Civil Northeast performed several operations simultaneously – above ground operations and underground excavation. Courtesy of Skanska USA Civil Northeast.

In 1990, the Metropolitan Transportation Authority approved a $425 million master plan for Grand Central Terminal, followed with an investment of $160 million in utility upgrades, Main Concourse improvements, and structural repairs. The former Main Waiting Room was also restored, becoming an exhibition and special events space in 1992. The city's Landmarks Preservation Commission made sure that this magnificent, historic building was not damaged.

GCA contractors did outstanding work on Grand Central Terminal. As James Moriarty, Jr., President, T. Moriarty and Son, and President of the GCA recalls: "We had a series of contracts with Metro-North and did projects over there for 10 to 12 years from the mid '80s through 1997. One of the projects involved installing four new 'chillers' on the roof. Chillers are large air conditioning units for a space that size. So we had 42nd St. closed on a weekend, using a 450 ton hydraulic crane to set new chillers on that beautiful limestone edifice. The Landmarks Commission was out there to make sure that nothing got damaged. And, thank God, it didn't. We were quite proud to be part of some of the redevelopment of Grand Central Terminal."[15]

As restoration and renovation continued, the rehabilitation generated more than 2,000 construction and construction related jobs. On October 1, 1998, the station was the setting for a Rededication Celebration of Grand Central Terminal. Throughout the nation and the entire world, the event was interpreted as one more indication that New York City had rebounded from the neglect and decay of earlier decades.

Grand Central Station is one of the city's many investments in its cultural treasures. With a four-year capital budget of over $1 billion, the city's Department of Cultural Affairs is the nation's largest funder of the arts. And, under the innovative "Percent for Art" program, started by Mayor Koch in 1982, one percent of the budget for city-funded construction projects is targeted to public artwork for municipal facilities.

FDR Drive (right)
Construction of temporary roadway
over East River.   Courtesy Skanska
USA Civil Northeast.

Bridges and Highways

After the collapse of part of the West Side Highway and instances of decaying infrastructure early in the 1970s, New York City has invested many more resources in maintaining its bridges and highways. Many bridges underwent major rehabilitation programs, and several were replaced, including the Shore parkway Bridge over Ocean Parkway in Brooklyn, the Third Avenue Bridge in Manhattan, and the Whitestone Expressway Bridge in Queens. In 2007, work began on the replacement of the Willis Avenue Bridge over the Harlem River.

Sanitation and Sewage

In one of the most important environmental advances in recent history, the North River Sewage Treatment Plant in Upper Manhattan opened in 1986. The treatment plant was built by GCA members Perini Corporation, Nicholas DiMenna & Sons, Inc., Poirier & McLane, Andrew Catapano Co., Inc., NAB Construction Corp., Slattery Construction Corp., Schiavone Construction, Corbetta Construction Co., and Terminal Construction Corp. Now that the facility is functioning, the sewers of Manhattan's West Side no longer flush directly into the Hudson River, as they had been doing since 1865.[16] Another new treatment plant opened in the Red Hook section of Brooklyn in 1987, ending the dumping of 40 million gallons a day of raw sewage into the East River from northwestern Brooklyn.

Recent federal regulations restrict the discharge of storm water runoff into the surface waters surrounding New York City. In order to comply with these rules, the New York State Department of Environmental Conservation ordered the city to construct a 30 million gallon underground storm surge storage tank for combined sewer runoff at Paerdegat Basin in Brooklyn which was built by GCA members John P. Picone, Inc., and Skanska Civil Northeast and a 43 million gallon holding tank in Flushing, Queens, which was built by E. E. Cruz and Company, Inc.

Meanwhile, a citywide facility for processing separately collected metal, glass, and plastic for recycling is being built on the South Brooklyn waterfront by the Hugo Neu Corporation, with completion scheduled in 2009.

On Staten Island, the huge Fresh Kills Landfill is being transformed into what will be one of the world's largest urban parks. Opened in 1948 and closed in 2001, the landfill's four mounds of accumulated trash are being sealed with plastic "geomembranes," eliminating odors and creating a safe and healthy environment. Together with these four hills, the surrounding wetlands and low-lying areas will comprise a 2,200-acre park—almost three times the size of Central Park. While the entire park will be completed by some time in the 2030s, one section leading from a neighborhood park to the Fresh Kills creek will open earlier. This part of the park will include a bike and walking path, a picnic area, and an observation tower. GCA member Tully Construction is working on this project.

City Water Tunnel No. 3 (right)

East Tunnel merging into main bellout.
Construction performed by GCA member
Schiavone Construction.
Photographer: Michael Patterson.
Courtesy Schiavone Construction.

Water Supply

As New York's population and economy have grown over the centuries, the need for a reliable supply of safe drinking water has been a recurring theme in the city's history. With increased funding from Mayor Bloomberg's administration, the progress has accelerated on City Water Tunnel 3 (which began during Mayor Koch's administration) the largest capital construction project in New York City's history. Connecting the city to its water supply system upstate, the tunnel will be more than 60 miles long. Work on the project began in 1970, and it is now expected to be completed in 2020.

In 2001, in order to comply with federal water quality standards, the U.S. Environmental Protection Administration ordered New York City to filter water from its Croton Aqueduct system, which currently supplies about 10 percent of the city's daily needs.[17] In response, the city broke ground for the Croton Filtration Plant in Van Cortlandt Park in the Bronx in May, 2005. GCA contractors Schiavone Construction Co. and Skanska USA Civil Northeast have been working on the project.

Freight Rail

In the only major investment in rail freight in New York City since World War II, a program is underway to allow railcars carrying trailers on flatcars, which are now standard equipment in the rest of the country, to be used in the New York area. In the past, this equipment could not be used because of bridges and other clearance restrictions. Substantially completed in 1997, the Full Freight Access Program includes track between New York City and Selkirk (just below Albany), as well as track and yards in Brooklyn, Queens, and the Bronx. The two most costly elements of the program are a rail freight yard in the South Bronx that was purchased and fitted for inter-modal transfer purposes (the Harlem River Rail yards, built between 1982 and 2006) and construction of a line along the Hudson River (the Oak Point Line, built between 1982 and 1998).

Oak Point Rail Link (below)

GCA member John Picone Inc. installs 36" x 1" steel caissons in the Harlem River using a barge mounted Manitowoc 4100 Ringer Crane. Caissons were used to support prestressed concrete beams and a poured-in-place concrete deck which carries a single rail line from North of Yankee Stadium to the Oak Point yard near the Triborough Bridge.

Staten Island once had freight rail service across the north shore and down the west shore. Purchased by the City from CSX in 1994, this line has been restored to serve the Howland Hook Marine Terminal and the new waste transfer station built at the former Fresh Kills landfill. The entire line reopened in 2006.

Port Projects

Continuing its tradition as a port, New York reopened the Red Hook Marine Terminal in Brooklyn in 1981 and the Howland Hook Marine terminal in Staten Island in 2000.

As a replacement for the passenger terminal on the Hudson River, the Economic Development Corporation developed a new passenger pier on the Brooklyn waterfront adjacent to the Red Hook Terminal in 2006.

Recovering from the Terrorist Attack

On September 11, 2001, New York City and the nation faced the ultimate test to their resilience. As almost the entire world knows, the World Trade Center was destroyed in the most devastating attack ever to take place on American soil. The attack killed nearly 3,000 people, destroyed 13.4 million square feet of office space,[18] forced 138,000 workers to relocate,[19] eliminated 83,000 jobs by the end of 2001, and reduced the city's tax revenues by $2 billion.[20]

Temporary PATH Station (top, left)
Courtesy Port Authority of New York and New Jersey.

Aerial View of Reconstruction of World Trade Center 2008 (bottom, left)
Courtesy Port Authority of New York and New Jersey.

But, as had happened in earlier crises, New York City rebounded from disaster—and the construction industry, including the contractors and the workers, responded to the challenge of trying to rescue the victims, remove the debris, and rebuild lower Manhattan.

Construction crews from the Grace Construction Company had been working on the West Side Highway and the George Washington Bridge. These workers were the first to reach the site where the Twin Towers of the World Trade Center has stood only a few hours earlier.[21] The City of New York mobilized dozens of other contractors to begin the work of clearing the debris.

Ground Zero was divided into four quadrants. Lead responsibility for each area was assigned to a different company—AMEC, Bovis Lend Lease, Tully, and Turner Construction. Starting soon after the attack, these crews would continue to work around the clock for months. The contractors' crews recovered 1.65 million tons of incinerated debris that they scooped up and hauled away.[22] These GCA members assisted in the cleanup effort: A.J. Pegno, Civetta Cousins, Conesco Doka, Ferrara Brothers, Gateway, Grace, Judlau, Koch Skanska, Moretrench, Nicholson, Weeks Marine, and Yonkers Contracting.

In the weeks that followed, it became clear that New York City's physical, social, and economic infrastructure all were holding up remarkably well.

When the huge, heavy towers collapsed, causing subsoil tremors equivalent to a small earthquake,[23] lower Manhattan's aged water, gas, and steam mains withstood the shock[24]—and the city was saved from a catastrophe even worse than it had already experienced.

Despite the hundreds of thousands of tons that fell on top of the area's subway system during the morning rush hour, not a single subway rider was killed. Indeed, the subway tunnels themselves, apart from a localized section directly below the towers, remained almost miraculously intact. The pioneering civil engineer William Barclay Parsons, his successors, and their construction crews had built better than anyone could have imagined.

After the attacks, every bridge and tunnel in the city was immediately closed. Nonetheless, even though the city had only a 24-hour supply of food, there was no looting of grocery stores. Nor did hoarders strip the shelves of food and emergency supplies. Somehow, the city's essential distribution systems stayed intact until the first trucks were allowed back across the George Washington Bridge on September 13th.

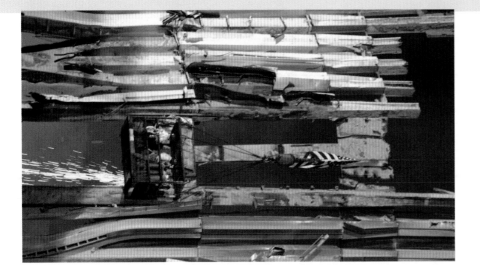

World Trade Center (right and below)

Many GCA members were involved in the recovery, clean-up, and eventual rebuilding at Ground Zero including Skanska Koch, Tully Construction, Beaver Concrete, Granite Construction, Skanska Civil Northeast, Kiewit Construction, Yonkers Contracting, Nicholson Construction, The Laquilla Group, and E.E. Cruz. Courtesy Skanska Koch.

Already suffering a slowdown from the collapse of the dot-com industry, the economy went into a downturn. But previously scheduled construction work continued, new projects were considered, and the city, the state, and the financial community pondered what to build where the Twin Towers once stood.

After nine months of nonstop effort, the cleanup of Ground Zero was completed ahead of schedule on May 30, 2002. By then, the reconstruction of the downtown infrastructure was well underway. A new 7 World Trade Center office building officially opened on May 23, 2006. A temporary PATH station opened on November 23, 2003, built by GCA members Yonkers Contracting Co., Tully Construction Co., and A.J. Pegno Construction Corp.

Meanwhile, the collapse of the World Trade Center damaged the IRT Broadway–Seventh Avenue Line (the 1 and 9 trains), which runs under the site. The trains were halted south of Chambers Street, re-routed until the tunnels, tracks and the station were repaired and reopened on September 15, 2002, ahead of schedule and under budget. The project was a joint venture of Tully Construction and AJ Pegno Construction.

But what would be built on Ground Zero? Coordinating the reconstruction, the Lower Manhattan Development Corporation, selected a plan including a 1776-foot tall Freedom Tower, with the height recalling the year that the United States declared its independence. On June 29, 2005, a design for the Freedom Tower was unveiled, with a conscious resemblance to the Twin Towers and new safety features. A memorial is also being built commemorating the events of September 11, 2001. The foundation for the memorial is being built by a joint venture of EE Cruz and Nicholson Construction. The site work is being performed by Phoenix Constructors, a quadruple venture which includes GCA members Skanska USA Civil Northeast, and Granite Construction.

Looking Forward

As in the previous century, public officials, business leaders, and organizations representing people from every walk of life, including construction craft workers are making bold plans to build an even greater New York. As John Donohoe, Chair of Moretrench Corp., explains: "The significant commitment New York City has made to infrastructure will serve the public well for decades to come. The city is investing in a third water tunnel, sewage treatment plants, the East Side Access Tunnel and the 2nd Avenue Subway. The potential for contracting in New York City today is tremendous."[25]

GCA contractors are doing their part to build New York City. Thomas Iovino, CEO and Founder, Judlau Contracting, reports: "Now we are working on the East Side Access Tunnel, which will connect the Long Island Railroad's Main and Port Washington lines in Queens to a new terminal underneath Grand Central Terminal in Manhattan. This new subway tunnel will go under the East River from Queens into the East Side of Manhattan."[26]

As the first decade of the twenty-first century concludes, the construction industry is very different from decades ago, especially in its use of technology and its emphasis on safety. Donohoe explains: "In the past, construction was much slower than today, not as well mechanized. Many of the pictures of work that took place in the 1920s and 30s show horse drawn trucks. When you look at the way the workers were dressed, they did not have safety equipment. The safety structures for protecting workers from falling did not really exist."

New technologies also increase efficiency, as Donohoe explains: "The greatest benefit that we see to technology is in communications. We can be in communication with the different crews that are working and with the trucks. We can be sure when they will arrive."

For contractors, the GCA is an essential element of doing business successfully. As Moriarty says: "The GCA is a very valuable asset for the city. The networking that you get for being a member has fostered a lot of relationships where we bid together with other firms, sometimes as the subcontractor with others or vice versa. Being a member of GCA brings a certain level of respect when you meet with the public agencies."[27]

Union labor is also indispensable. Iovino notes: "Union members are the most reliable workforce in New York. Not just reliable but high-quality. They really bring value to the work that they perform, and they take pride in their work. I'm very proud of the people that work for me that are in the unions."[28] Adds Moriarty: "Union members work safer because of the amount of training they receive that I don't believe the non-union guys get, unfortunately for them. Union members do their work in a timely and professional manner."

Indeed, today's construction workers may be even more productive than their predecessors. Thomas King, a past president of the GCA who recently retired from Schiavone Construction Company, observes: "Young workers are more home types. People go home at night, instead of going to the bars."

New York City's extraordinary needs, resources, traditions, and workforce have created a construction industry that is preparing to meet the challenges of the twenty-first century. As Iovino concludes: "There is a lot of competition. New York City demands the best and gets the best. Because of the competition, the industry keeps getting better. If you don't improve in what you do, you're not going to stay in business."

Ground Zero is being rebuilt, along with every facet of all five boroughs. New York is known as the safest big city in this country, an economic center nationally and internationally, and an incubator and a magnet for energetic and ambitious people, from all 50 states, and the entire world. And with PlaNYC, Mayor Michael Bloomberg's plan for environmentally sustainable growth, there is a pathway to the future that includes accommodating a million more people, modernizing and maintaining the city's infrastructure, and providing for economic growth that is sustainable. Working together, we will create a greener, greater New York.

New Yorkers and all Americans have new reason to admire the soaring spirit of the nation's greatest city, with its infinite capacity to dream, to build, and to recover from adversity. This is the story of the past 100 years of building and rebuilding New York City. The next century's story promises to be even more inspiring.

Coney Island Sewage Treatment Plant (above, right)
Contract 15G. Courtesy Silverite Construction.

Reconstruction of BQE (above, left)
Brooklyn Queens Expressway. Deck removal and replacement from Tillary Street to the Brooklyn Navy Yard. Courtesy Defoe Corporation.

JetBlue at JFK Airport (opposite page)
Delaney Associates LP installing force main sewer at JFK Airport in Queens, New York. June 28, 2007. Courtesy Delaney Associates LP.

Asphalt Paving

Legend has it that immigrants believed that, in New York City, "The streets were paved with gold." But, in reality, until the late nineteenth century, the city's streets were paved with bricks, cobblestones, or even, as recently as 1905, with wooden blocks.

Compared to these early urban pavements, concrete paving represented a significant advance. But, while concrete was an inexpensive material for general construction purposes, it was still relatively expensive to use in the volume required to pave city streets. To add to the difficulties, when used on roads, concrete is subject to cracking and crumbling and is relatively expensive to maintain.

By the beginning of the twentieth century, New York City was requiring contractors to use a new combination of materials to pave most roads. Usually, the base layer would be concrete. But the pavement would use asphalt as the "glue" for "aggregate particles," such as sand, crushed stone, and asphalt bits.

Asphalt—the thick and sticky residue from refining crude oils—is ideally suited to paving roads. But it took builders much of the nineteenth century to discover its uses. Back in 1816, the Scottish engineer John McAdam pioneered the use of crushed stone for road-building, binding the stones together with coal tar. Eventually, the crushed stone pavements came to be called "macadam," and the tar was called "tarmac."

Grand Central Parkway Extension (left)
View west from centerline of parkway under the 78 St. bridge showing paving operations for roadway slab. October 10, 1935. Photographer unknown. Courtesy MTA Bridges and Tunnels Special Archive.

During the following decades, cars carrying asphalt rocks over macadam roads sometimes dropped the material accidentally onto the crushed stone. The sun would melt the asphalt, or passing vehicles would crush it into a smooth, solid surface layer that bound the stones together. This discovery created a demand for asphalt, and manufacturers learned how to refine it from petroleum, creating an inexpensive and useful paving material.

In New York City, one of the first high-speed roads to use asphalt paving on top of a concrete base was the Bronx River Parkway, which began in 1901 and completed in 1923. In 1922, when the parkway commissioners decided to use asphalt, one consideration was that the asphalt would cover the construction joints in the concrete, thus producing a smooth and quiet ride. Another factor was that asphalt would require less maintenance than other kinds of surfacing materials. Yet another factor was aesthetic: The parkway commissioners were concerned that a concrete road "would stand out as a starkly artificial intrusion in the parkway's naturalistic landscape." But they believed that a dark road surface like asphalt would blend in better with the surroundings.

Almost a decade and a half later, the city achieved another innovation in the use of asphalt. Opening in 1936, the Harlem River lift span of the Triborough Bridge avoided the weight of a concrete roadway by using asphalt planks on top of steel-plate road deck girders.

Grand Central Parkway Extension (right)
View, east from centerline of parkway and 78 St. bridge showing fine grading for parkway pavement and pavement forms set. October 10, 1935. Photographer unknown. Courtesy MTA Bridges and Tunnels Special Arch ve.

Mixing and Making Asphalt

In the early days of asphalt paving, molten asphalt was heated in kettles. Then it was sprinkled onto a pre-rolled surface of sand or crushed stones. Then the roadway was compacted by hand or by a horse-drawn or steam-powered roller. Gradually, it became clear that mixing the stone and asphalt together before laying it provided a superior result. The crushed rock was dried on shallow iron trays heated over open coal fires, hot asphalt was poured on top, and the mixture was stirred with a shovel until all the aggregate was coated. The hot mixture was then dumped on the street to be spread and smoothed with shovels, rakes, brooms, squeegees, and tampers. When the asphalt was in place, a roller ran over it.

Such labor intensive methods were used well into the 1930s, even though specialized machinery had been developed for paving with concrete, which could be mixed at room temperature. The first drum mixers for asphalt—made by modifying concrete mixers—were used in 1910. In that year, the borough of Brooklyn bought a Lutz machine to produce hot asphalt for resurfacing its streets, using the "heater method." Five years later, Manhattan followed suit. In the late 1920s, automated gasoline-powered equipment—again, modified from concrete machinery—began to be used for laying asphalt. In the 1930s, vibrating screens and pressure-injection systems began to be used in mixing plants, and a number of improved paving machines designed specifically for asphalt were developed.

Like ready-mix concrete, hot-mixed asphalt cannot be transported for too long a period of time before it becomes unusable. Therefore, asphalt plants must be located reasonably close to population centers.

In order to control its own supply of this essential, high-volume construction material, New York City began building its own asphalt plants early in the twentieth century. The city's first asphalt plant was built in Brooklyn in 1907 on the Gowanus Creek at 6th Street (not far from where the Cranford Company would later locate its concrete plant).

New Asphalt Plant (left)
In the late 1980s, NAB construction built a new asphalt plant for the city, which is capable of producing up to 300 tons of asphalt per hour. Courtesy NAB.

Van Wyck Expressway (above)
Camera at 133rd Avenue Bridge, facing south. June 26, 1951.
Photographer: Roy Foody for New York State Department of Public Works.
Courtesy MTA Bridges and Tunnels Special Archive.

The savings in asphalt procured from private sources led the city to begin developing new plants in other boroughs. In 1939, in order to eliminate the smoke, dust, and odor nuisances that the Brooklyn plant had caused, the city began building a modern "streamlined" plant in Manhattan, on 90th Street and the East River, just a few blocks north of the mayor's official resicence, Gracie Mansion. This plant opened in 1945 and the highway department gave it to the parks department in 1970. Where the asphalt plant once stood, there now is a park and recreation area called Asphalt Green.

More modern (but less attractive) municipally owned asphalt plants have been built in more recent decades.

"What GCA does for us is to help us to be as sophisticated as any other contractor in New York City using the same unions because they negotiate and implement the collective bargaining agreement. The GCA is always there for us."

Bruce Junge, *Executive Vice President, Beaver Concrete Construction Co., Inc.*

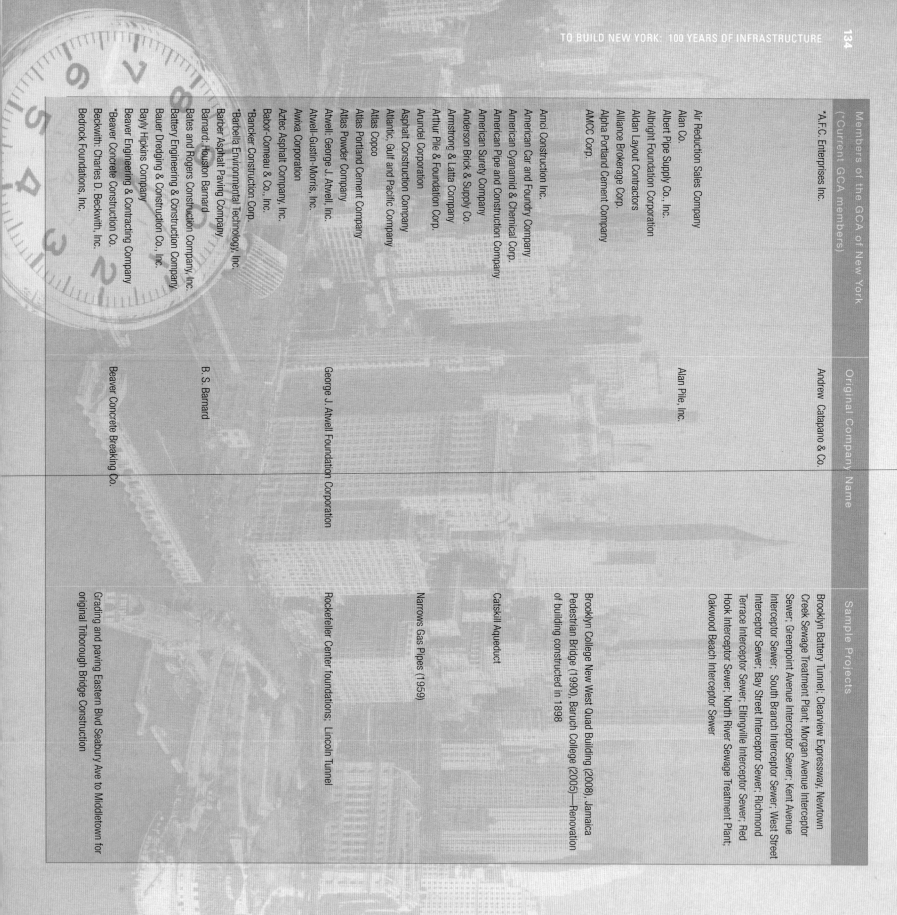

Members of the GCA of New York (*Current GCA members)	Original Company Name	Sample Projects
*A.F.C. Enterprises Inc.	Andrew Catapano & Co.	Brooklyn Battery Tunnel; Clearview Expressway; Newtown Creek Sewage Treatment Plant; Morgan Avenue Interceptor Sewer; Greenpoint Avenue Interceptor Sewer; Kent Avenue Interceptor Sewer; South Branch Interceptor Sewer; West Street Interceptor Sewer; Bay Street Interceptor Sewer; Richmond Terrace Interceptor Sewer; Eltingville Interceptor Sewer; Red Hook Interceptor Sewer; North River Sewage Treatment Plant; Oakwood Beach Interceptor Sewer
Air Reduction Sales Company		
Alan Co.		
Albert Pipe Supply Co., Inc.		
Albright Foundation Corporation		
Aldan Layout Contractors		
Alliance Brokerage Corp.		
Alpha Portland Cement Company		
AMCC Corp.		
Amci Construction Inc.	Alan Pile, Inc.	Brooklyn College New West Quad Building (2008), Jamaica Pedestrian Bridge (1990), Baruch College (2005)—Renovation of building constructed in 1898
American Car and Foundry Company		
American Cyanamid & Chemical Corp.		
American Pipe and Construction Company		
American Surety Company		Catskill Aqueduct
Anderson Brick & Supply Co.		
Armstrong & Latta Company		
Arthur Pile & Foundation Corp.		
Arundel Corporation		
Asphalt Construction Company		
Atlantic, Gulf and Pacific Company		Narrows Gas Pipes (1959)
Atlas Copco		
Atlas Portland Cement Company		
Atlas Powder Company		
Atwell, George J. Atwell, Inc.	George J. Atwell Foundation Corporation	Rockefeller Center foundations; Lincoln Tunnel
Atwell-Gustin-Morris, Inc.		
Awixa Corporation		
Aztec Asphalt Company, Inc.		
Babor-Comeau & Co., Inc.		
*Bancker Construction Corp.		
*Barbella Environmental Technology, Inc.		
Barber Asphalt Paving Company		
Barnard, Houston Barnard	B. S. Barnard	
Bates and Rogers Construction Company, Inc.		
Battery Engineering & Construction Company		
Bauer Dredging & Construction Co., Inc.		
Bayly Hipkins Company		
Beaver Engineering & Contracting Company		
*Beaver Concrete Construction Co.	Beaver Concrete Breaking Co.	Grading and paving Eastern Blvd Seabury Ave to Middletown for original Triborough Bridge Construction
Beckwith; Charles D. Beckwith, Inc.		
Bedrock Foundations, Inc.		

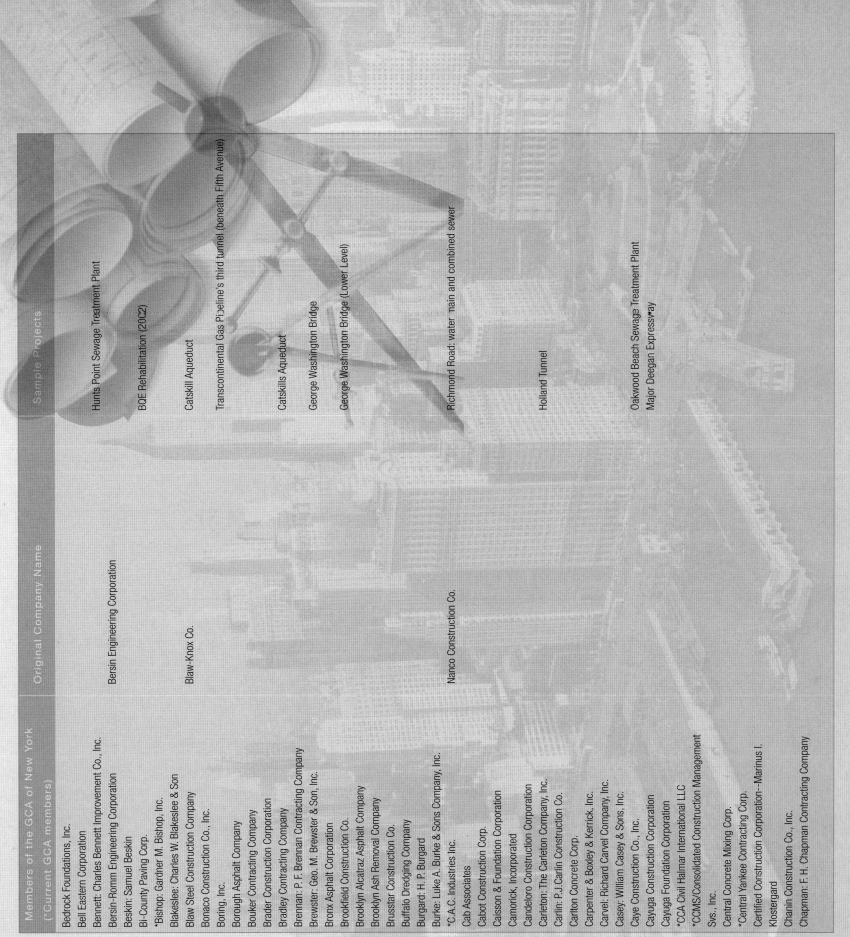

Members of the GCA of New York (*Current GCA members)	Original Company Name	Sample Projects
Bedrock Foundations, Inc.		
Bell Eastern Corporation		
Bennett: Charles Bennett Improvement Co., Inc.	Bersin Engineering Corporation	Hunts Point Sewage Treatment Plant
Bersin-Romm Engineering Corporation		
Beskin: Samuel Beskin		
Bi-County Paving Corp.		BQE Rehabilitation (20C2)
*Bishop: Gardner M. Bishop, Inc.		
Blakeslee: Charles W. Blakeslee & Son		Catskill Aqueduct
Blaw Steel Construction Company	Blaw-Knox Co.	
Bonaco Construction Co., Inc.		Transcontinental Gas Pipeline's third tunnel (beneath Fifth Avenue)
Boring, Inc.		
Borough Asphalt Company		Catskills Aqueduct
Bouker Contracting Company		
Brader Construction Corporation		George Washington Bridge
Bradley Contracting Company		
Brennan: P. F. Brennan Contracting Company		George Washington Bridge (Lower Level)
Brewster: Geo. M. Brewster & Son, Inc.		
Bronx Asphalt Corporation		
Brookfield Construction Co.		
Brooklyn Alcatraz Asphalt Company		
Brooklyn Ash Removal Company		
Brusstar Construction Co.		Richmond Road: water main and combined sewer
Buffalo Dredging Company		
Burgard: H. P. Burgard		
Burke: Luke A. Burke & Sons Company, Inc.		
*C.A.C. Industries Inc.	Nanco Construction Co.	
Cab Associates		
Cabot Construction Corp.		Holland Tunnel
Caisson & Foundation Corporation		
Camorick, Incorporated		
Candeloro Construction Corporation		
Carleton: The Carleton Company, Inc.		
Carlin: P.J.Carlin Construction Co.		
Carlton Concrete Corp.		Oakwood Beach Sewage Treatment Plant
Carpenter & Boxley & Kerrick, Inc.		Major Deegan Expressway
Carvel: Richard Carvel Company, Inc.		
Casey: William Casey & Sons, Inc.		
Caye Construction Co., Inc.		
Cayuga Construction Corporation		
Cayuga Foundation Corporation		
*CCA Civil Halmar International LLC		
*CCMS/Consolidated Construction Management Svs., Inc.		
Central Concrete Mixing Corp.		
*Central Yankee Contracting Corp.		
Certified Construction Corporation--Marinus I. Klostergard		
Chanin Construction Co., Inc.		
Chapman: F. H. Chapman Contracting Company		

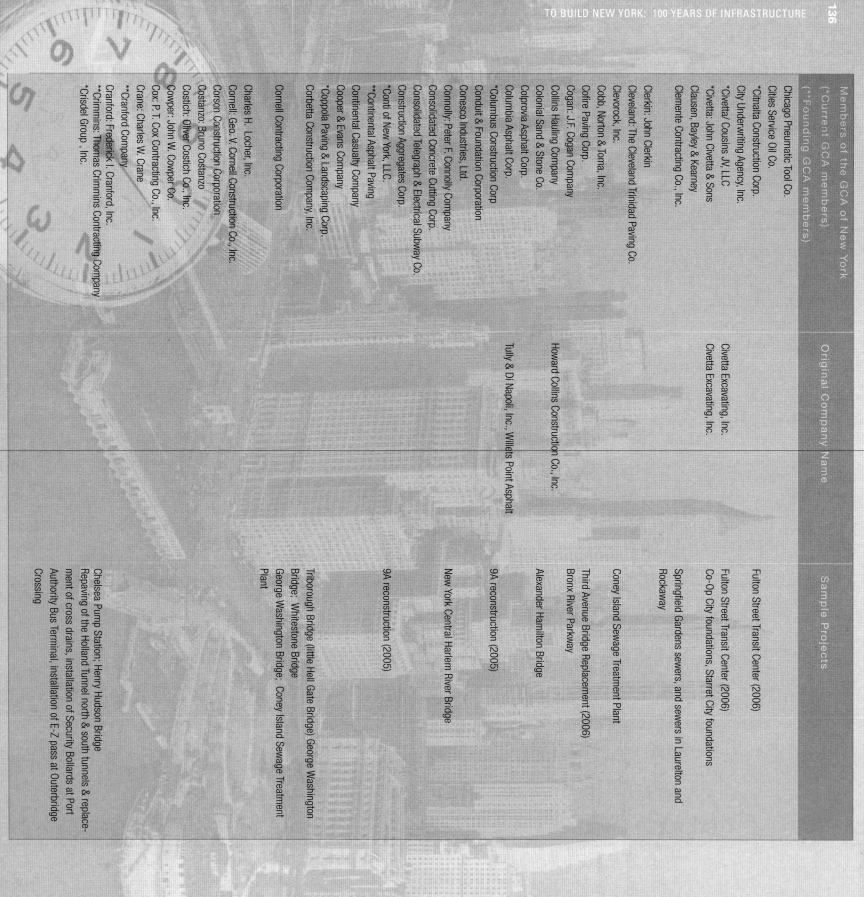

Members of the GCA of New York

(*Current GCA members)

(**Founding GCA members)

	Original Company Name	Sample Projects
Chicago Pneumatic Tool Co.		
Cities Service Oil Co.		
*Citnalta Construction Corp.		
City Underwriting Agency, Inc.		
*Civetta/ Cousins JV, LLC		
*Civetta: John Civetta & Sons	Civetta Excavating, Inc.	Fulton Street Transit Center (2006)
Clausen, Bayley & Kearney		
Clemente Contracting Co., Inc.	Civetta Excavating, Inc.	Fulton Street Transit Center (2006)
		Co-Op City foundations; Starret City foundations
Clerkin: John Clerkin		
Cleveland: The Cleveland Trinidad Paving Co.		Springfield Gardens sewers, and sewers in Laurelton and Rockaway
Clevorock, Inc.		
Cobb, Norton & Toma, Inc.		Coney Island Sewage Treatment Plant
Cofire Paving Corp.		
Cogan: J.F. Cogan Company		Third Avenue Bridge Replacement (2006)
Collins Hauling Company	Howard Collins Construction Co., Inc.	Bronx River Parkway
Colonial Sand & Stone Co.		
Coprovia Asphalt Corp.		Alexander Hamilton Bridge
Columbia Asphalt Corp.		
*Columbus Construction Corp		
Conduit & Foundation Corporation		9A reconstruction (2005)
Conesco Industries, Ltd.		
Connolly: Peter F. Connolly Company		
Consolidated Concrete Cutting Corp.		New York Central Harlem River Bridge
Consolidated Telegraph & Electrical Subway Co.		
Construction Aggregates Corp.		
*Conti of New York, LLC.	Tully & Di Napoli, Inc., Willets Point Asphalt	9A reconstruction (2005)
**Continental Asphalt Paving		
Continental Casualty Company		
Cooper & Evans Company		
*Coppola Paving & Landscaping Corp.		
Corbetta Construction Company, Inc.		Triborough Bridge (little Hell Gate Bridge) George Washington Bridge; Whitestone Bridge
		George Washington Bridge; Coney Island Sewage Treatment Plant
Cornell Contracting Corporation		
Charles H. Locher, Inc.		
Cornell: Geo. V. Cornell Construction Co., Inc.		
Corson Construction Corporation		
Costanzo: Bruno Costanzo		
Costich: Oliver Costich Co., Inc.		
Cowper: John W. Cowper Co.		
Cox: P. T. Cox Contracting Co., Inc.		
Crane: Charles W. Crane		
**Cranford Company		Chelsea Pump Station; Henry Hudson Bridge
Cranford: Frederick I. Cranford, Inc.		Repaving of the Holland Tunnel north & south tunnels & replacement of cross drains, installation of Security Bollards at Port Authority Bus Terminal, installation of E-Z pass at Outerbridge Crossing
**Crimmins: Thomas Crimmins Contracting Company		
*Crisdel Group , Inc.		

Members of the GCA of New York (*Current GCA members) (**Founding GCA members)	Original Company Name	Sample Projects
*Cruz Construction Corp.		Oak Beach Interceptor Sewer
*Cruz: E.E. Cruz & Co., Inc.		World Trade Center Transportation Hub in joint venture with Nicholson Construction, World Trade Center Memorial Foundation, Cut-and-cover tunnel at JFK Airport under main taxiways, Flushing Bay CSO Retention Facility Phase III and IV
* Cruz Enterprises, LLC		
Cummins Diesel		
Cummins, Coakley & Booth, Inc.		
Cunningham Asphalt Construction Co., Inc.		
Curly Construction Co., Inc.		
Curtis: Harry F. Curtis		
*D. Gangi Contracting Corp.		
Dale & Rankin Company, Inc.		
*Danella Construction of NY, Inc.		
Daniels: Oscar Daniels Company		
Davis Construction Corporation		Delaware aqueduct (Frazer-Davis Construction Company)
de Stefano: Nicholas de Stefano		
*Deboe Construction Corp.		
Deepwater Contracting Co., Inc.		
*Defoe Corp.		Cross Bronx Expressway;Bruckner Interchange renovation, Brooklyn Queens Expressway Park Avenue rehabilitation, Gowanus / Connector Ramp
		Steinway Tunnels (IRT)
**Degnon Contracting Company		Major Deegan Expressway; George Washington Bridge (Lower Level)
Del Balso Construction Corporation		
*Delaney Associates LP		
Delee Contracting Co., Inc.		
Delhan Construction Company, Inc.		
*Delma Construction Co. Inc.	Delma Engineering Corp.	
Di Marco & Reimann, Inc.		
Di Menna: Jerome Di Menna Co., Inc.		
Di Menna: Nicholas Di Menna & Sons Construction Corp.	Remo Engineering, Nicholas Di Menna & Sons, Inc.	Coney Island Sewage Treatment Plant; North River Sewage Treatment Plant; sewers utilities, and watermains for co-op city; East River drive from 1939 to 1941; Restoration of Harlem River Seawall from Dyckman Street to Washington Bridge; Restoration of Roosevelt Island Seawall West Channel of the East River
*Diamond Asphalt Corp.		
Diebitsch: Emil Diebitsch		
Dienst: A. P. Dienst Company		
Dietz-Fallert Construction Company		
Digram & Company, Inc.		
Dippel: William T. Dippel, Inc.		
Dittmar Powder Works, Inc.		
Dixon: L. E. Dixon & Company		
Dock Contractor Company		
Donovan: John E. Donovan		
*Dragados USA, Inc.		
Dravo Corporation		Delaware Aqueduct
Drilled-in Caisson Corporation		

Members of the GCA of New York
(*Current GCA members)
(**Founding GCA members)

	Original Company Name	Sample Projects
Driscoll: George F. Driscoll Co.		
*Dryden Diving Co. Inc.		
*D–Star Waterproofers, Inc.		
Du Pont: E. I. Du Pont de Nemours & Co., Inc.		
Duffy: Edward J. Duffy Company	Duffy Construction Corporation	Reconstruction of Kensico Dam, Muscoot Dam, and Kensico Spillway Bridge; Catskill UV Plant (2008), 65th Street Rail Yard Transfer Bridges; Design/Build Hudson Line Stations; Reconstruction Cross Bronx Expressway; Reconstruction Harlem River Drive; LIE Westbound Collector Distributor
Duffy: J. P. Duffy Company		
Duncan: John H. Duncan, Inc.		
E. A. Hauling Co.		
Earth: The Earth Bank Co., Inc.		
Eastern Contracting Company		
*Eberhart Construction Co. Inc.		
*Ecco III Enterprises, Inc.		
		Verrazano Bridge Lower Level Deck Replacement, Lily Pond Avenue Bridge— Structural Steel and Deck Replacement, Cross Bay Veterans Bridge—Deck and Structural Rehabilitation, Gowanus Expressway Widening and Deck Replacement, East River Bridges preventive maintenance and repairs at various locations.
Edenwald Contracting Co.		
Edgecliff Construction Corp.		
Edwards & Flood, Inc.		
*El Sol Contracting & Construction. Corp.		
Electric Welding Company of America		
Elmhurst Contracting Co., Inc.		
Emerson. Hines, Inc.		
Empire Engineering Co., Inc.	The Empire Construction Company	Throgs Neck Bridge; Newtown Creek Sewage Treatment Plant
Equipco		
Erickson Equipment Co., Inc.		
*F&S Contracting LLC.		
Faircroft Engineering Corporation		
Fajella: Francis J. Fajella & Co., Inc.		
*Falco Construction Corp.		
Farley: James A. Farley & Company, Inc.		
Farrier: The W. W. Farrier Co.		
Federal Signal Company		
Federation Bank and Trust Company		
Federico: A. Federico		
Fehlhaber Pile Co., Inc.		
*Ferreira Construction Co. Inc.		
Fidelity and Deposit Company		
First National City Bank		
Fitzgerald: W. J. Fitzgerald		Holland Tunnel
Fitzpatrick: Edward B. Fitzpatrick Jr. Associates Inc.		
*Fleet Trucking Inc.		
Flinn–O'Rourke Company, Inc.	George H. Flinn Corporation	Grand Central Station; Coney Island Sewage Treatment Plant
Foley Brothers, Inc.		George Washington Bridge; Brooklyn-Battery Tunnel

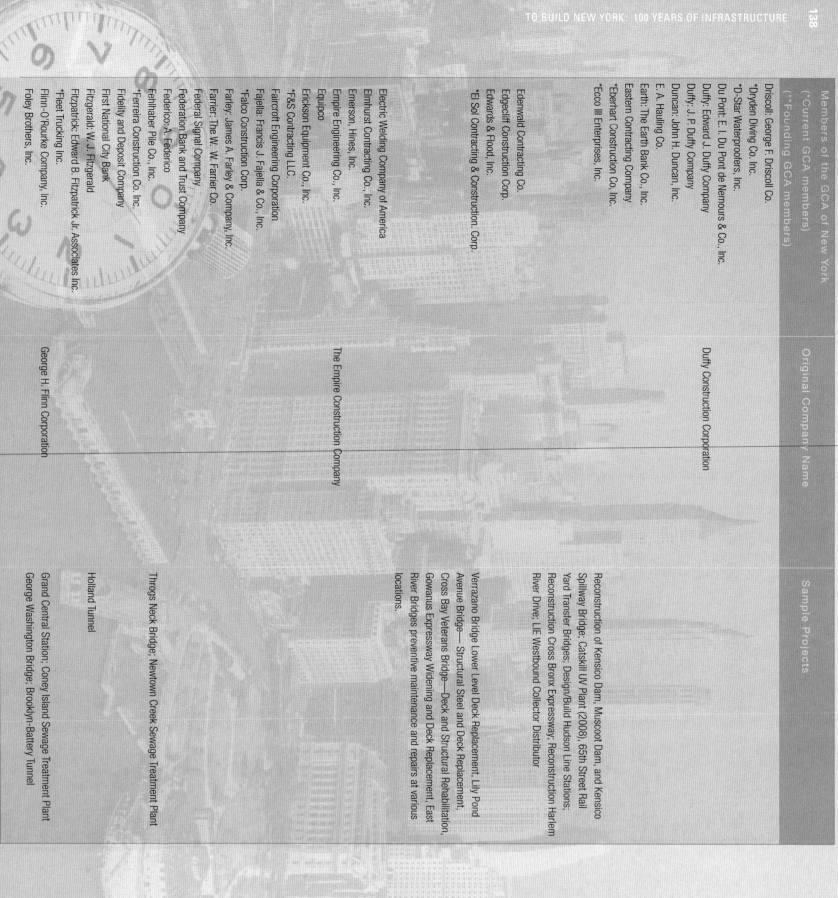

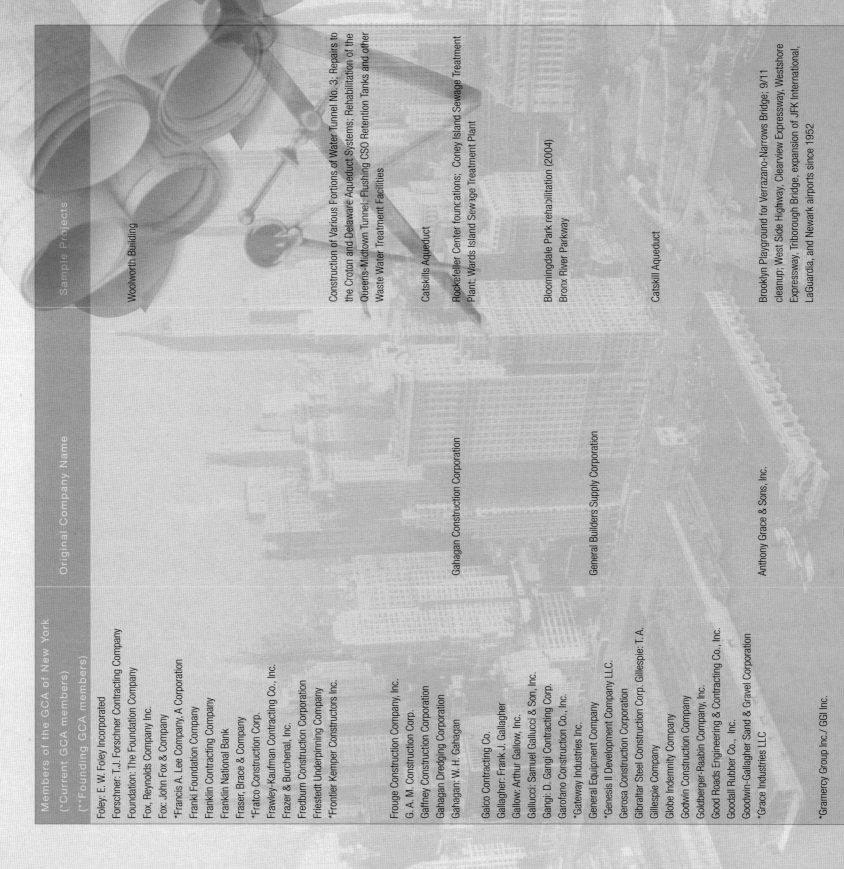

Members of the GCA of New York

(*Current GCA members)

(**Founding GCA members)

Original Company Name	Sample Projects
Foley: E. W. Foley Incorporated	
Forschner: T.J. Forschner Contracting Company	
Foundation: The Foundation Company	Woolworth Building
Fox, Reynolds Company Inc.	
Fox: John Fox & Company	
*Francis A. Lee Company, A Corporation	
Franki Foundation Company	
Franklin Contracting Company	
Franklin National Bank	
Fraser, Brace & Company	
*Fratco Construction Corp.	
Frawley-Kaufman Contracting Co., Inc.	
Frazer & Burchenal, Inc.	
Fredburn Construction Corporation	
Friestedt Underpinning Company	
*Frontier Kemper Constructors Inc.	Construction of Various Portions of Water Tunnel No. 3; Repairs to the Croton and Delaware Aqueduct Systems; Rehabilitation of the Queens-Midtown Tunnel; Flushing CSO Retention Tanks and other Waste Water Treatment Facilities
Frouge Construction Company, Inc.	
G. A. M. Construction Corp.	
Gaffney Construction Corporation	Catskills Aqueduct
Gahagan Dredging Corporation	
Gahagan: W. H. Gahagan	Gahagan Construction Corporation	Rockefeller Center foundations; Coney Island Sewage Treatment Plant; Wards Island Sewage Treatment Plant
Galco Contracting Co.	
Gallagher: Frank J. Gallagher	
Gallow: Arthur Gallow, Inc.	
Gallucci: Samuel Gallucci & Son, Inc.	
Gangi: D. Gangi Contracting Corp.	Bloomingdale Park rehabilitation (2004)
Garofano Construction Co., Inc.	Bronx River Parkway
*Gateway Industries Inc.	
General Equipment Company	General Builders Supply Corporation
*Genesis II Development Company LLC.	
Gerosa Construction Corporation	
Gibraltar Steel Construction Corp. Gillespie: T. A. Gillespie Company	Catskill Aqueduct
Globe Indemnity Company	
Godwin Construction Company	
Goldberger-Raabin Company, Inc.	
Good Roads Engineering & Contracting Co., Inc.	
Goodall Rubber Co., Inc.	
Goodwin-Gallagher Sand & Gravel Corporation	
*Grace Industries LLC	Anthony Grace & Sons, Inc.	Brooklyn Playground for Verrazano-Narrows Bridge; 9/11 cleanup; West Side Highway, Clearview Expressway, Westshore Expressway, Triborough Bridge, expansion of JFK International, LaGuardia, and Newark airports since 1952
*Gramercy Group Inc./ GGI Inc.	

Members of the GCA of New York
(*Current GCA members)
(**Founding GCA members)

	Original Company Name	Sample Projects
*Granite Construction Northeast, Inc.	Granite Halmar	Taconic Parkway Reconstruction, 2008; American Airlines, JFK Landside & Airside Reconstruction, 2007; Metro North Third Track, 2004; Howland Hook Marine Terminal, 2004; Stillwell Avenue Terminal Reconstruction 2005; Belt Parkway Bridge & Ocean Parkway Interchange Reconstruction 2005; South Ferry Terminal Subway Station (JV), 2007.; Signal System Modernization, Concourse Line, 2007.; Grand Avenue Bus Depot & Maintenance Facility 2008; Amtrak Vent Shafts, ongoing; 53rd & 6th /Park Avenue Vent Plants, ongoing; Intercounty Connector-MA, ongoing; WTC Transit Hub Replacement. Transcontinental Gas Pipe Line Corporation's third underwater pipeline (1958)
Great Lakes Dredge & Dock Company		Brooklyn Bridge Park (2001); Owls Head Sewage Treatment Plant beautification
Gregory: Louis D. Gregory		
Gregory: The Gregory Contracting Co., Inc.		
Griffin Wellpoint Corporation		
Grimmer: Chas. Grimmer & Son		
Gross: William A. Gross Construction Associates, Inc.		
Groves: S. J. Groves & Sons Company		
Grow Construction Co. Inc.		63rd Street tunnel; Clearview Expressway Brooklyn-Battery Tunnel; Oakwood Beach Interceptor Sewer; Newtown Creek Sewage Treatment Plant; Morgan Avenue Interceptor Sewer; Greenpoint Avenue Interceptor Sewer; Kent Avenue Interceptor Sewer; South Branch Interceptor Sewer; West Street Interceptor Sewer; Bay Street Interceptor Sewer; Richmond Terrace Interceptor Sewer; Eltingville Interceptor Sewer; Red Hook Interceptor Sewer; Water Tunnel No. 3; Alexander Hamilton Bridge
Guise: John C. Guise, Inc.		
Gull Contracting Co., Inc.	GCC, Inc., Mac Asphalt	Major Deegan Expressway; Brooklyn-Battery Tunnel; George Washington Bridge (Martha); Hunts Point Terminal Market
H. & P. Excavation & Foundation Co.		
Hallen: The Hallen Company, Inc.	Harry S. Hart, Inc.	
*Halmar International LLC.		Yankee Stadium Train Station, WTC-Tower 2, JFK British Airways Terminal, JFK International Arrivals Terminal, Grand Central Parkway reconstruction.
Hardaway Contracting Co.		
Harlem Contracting Company		
Hart & Early Co., Inc.		
Hart: John J. Hart		
Hartford Accident and Indemnity Company		
Hassam Paving Company		
Hastings Pavement Company		
*Hayward Baker Inc.	The Hayward Company	
Hazell: A. M. Hazell, Inc.		
Healy: S. A. Healy Company		East River Drive (seawall and fill), Delaware Aqueduct
Heller: Tobias Heller & Son, Inc.		
Helmers: Nick F. Helmers, Inc.		
Hendrickson Bros., Inc.		Northern State Parkway
Hercules Powder Company		

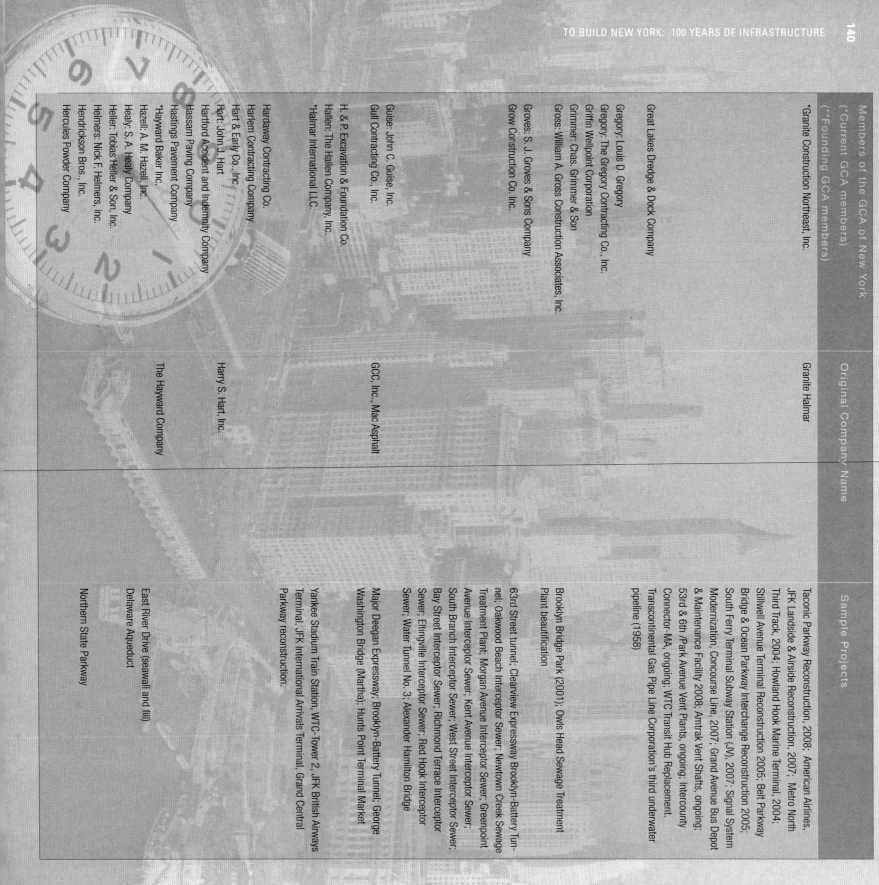

Members of the GCA of New York (*Current GCA members) (**Founding GCA members)	Original Company Name	Sample Projects
Heydt-Mugler Company, Inc.		
Heyman: The Heyman & Goodman Co., Inc.		
*HHJR Construction LTD.		
D/B/A/ Accurate Enterprises		
*HHM Associates Inc.		
Highway Improvement & Repair Co., Inc.		
Holbrook, Cabot & Rollins Corporation		
Holler & Shepard		
Holler: John M. Holler		
Horn Construction Co., Inc.		Throgs Neck Bridge; Newtown Creek Sewage Treatment Plant; Bruckner Boulevard Bridge (replacement, 1950); George Washington Bridge (Martha; pouring concrete on roadway)
Horton: James R. Horton		
Hudson Contracting Co., Inc.		
Hudson River Stone Corporation		
Hughes Bros., Inc.		
Hunterspoint Lumber & Supply Co., Inc.		
*Hycon Construction Systems Corp.		
Interboro Surface Co., Inc.		
Inter-Continental Construction Corporation		
International Underwater Contractors, Inc.		
*Interstate Payroll Company		
Iron Trap Rock Company		
*Island Foundations Corp.		
*J. D'Annunzio & Sons Inc.		
*J.T. Clearly, Inc.		
James McCullagh Co., Inc.		
Jarrett-Chambers Company, Inc.		
*JCI (Joy Contractors, Inc.)		
Jemko Construction Corp.		
Jenkins: J. W. Jenkins, Inc.		
Jersey City Dry Docks Co.		
*JMA Concrete Construction Co., Inc.		
Jobson-Gifford Company		
*John P. Picone Inc.		
Johnson, Drake & Piper, Inc.		Whitestone Bridge; Major Deegan Expressway; Verrazano-Narrows Bridge
Johnson: H. Johnson & Son, Inc.		
Johnson: Joseph Johnson's Sons		Verrazano-Narrows Bridge (Anchorages), Lincoln Tunnel
Joy Manufacturing Company	Arthur A. Johnson Corp.	

Hygrade Builders Supply Company, Inc.
Icanda Limited
Icos Corporation of America — World Trade Center foundation
Igoe Brothers
Industries Development Corporation
Ingersoll-Rand Company — Howland Hook Freight Rail Line; Staten Island Railroad Line
Ingram & Greene, Inc. — Visy Paper Mill
*Inspectronic Corporation
*Integrated Structures Corp.

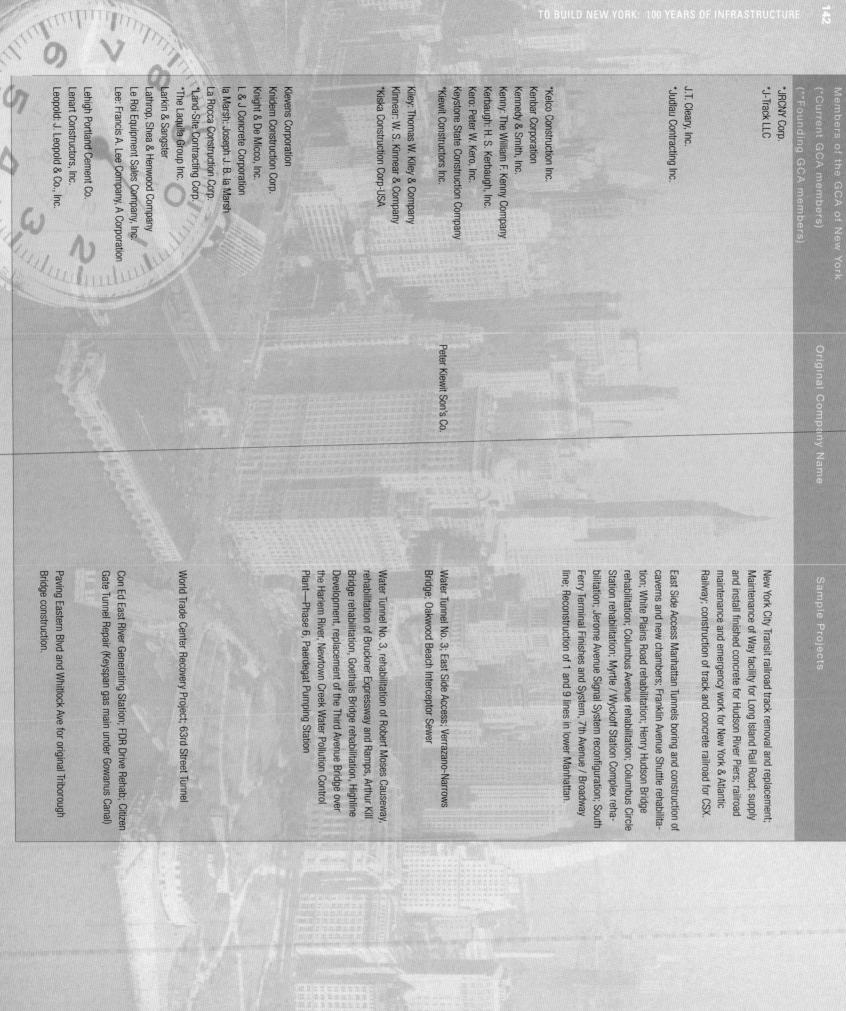

Members of the GCA of New York
(*Current GCA members)
(**Founding GCA members)

Members of the GCA of New York	Original Company Name	Sample Projects
*JRCNY Corp. *J-Track LLC		New York City Transit railroad track removal and replacement; Maintenance of Way facility for Long Island Rail Road; supply and install finished concrete for Hudson River Piers; railroad maintenance and emergency work for New York & Atlantic Railway; construction of track and concrete railroad for CSX.
J.T. Cleary, Inc. *Judlau Contracting Inc.		East Side Access Manhattan Tunnels boring and construction of caverns and new chambers; Franklin Avenue Shuttle rehabilitation; White Plains Road rehabilitation; Henry Hudson Bridge rehabilitation; Columbus Avenue rehabilitation; Columbus Circle Station rehabilitation; Myrtle / Wyckoff Station Complex rehabilitation; Jerome Avenue Signal System reconfiguration; South Ferry Terminal Finishes and System, 7th Avenue / Broadway line; Reconstruction of 1 and 9 lines in lower Manhattan.
*Kelco Construction Inc. Kenbar Corporation Kennedy & Smith, Inc. Kenny: The William F. Kenny Company Kerbaugh: H. S. Kerbaugh, Inc. Kero: Peter W. Kero, Inc. Keystone State Construction Company *Kiewit Constructors Inc.	Peter Kiewit Son's Co.	Water Tunnel No. 3: East Side Access; Verrazano-Narrows Bridge; Oakwood Beach Interceptor Sewer
Kiley: Thomas W. Kiley & Company Kinnear: W. S. Kinnear & Company *Kiska Construction Corp-USA		Water Tunnel No. 3, rehabilitation of Robert Moses Causeway, rehabilitation of Bruckner Expressway and Ramps, Arthur Kill Bridge rehabilitation, Goethals Bridge rehabilitation, Highline Development, replacement of the Third Avenue Bridge over the Harlem River, Newtown Creek Water Pollution Control Plant—Phase 6, Paerdegat Pumping Station
Klevens Corporation Knidem Construction Corp. Knight & De Micco, Inc. L & J Concrete Corporation la Marsh: Joseph J. B. la Marsh La Rocca Construction Corp. *Land-Site Contracting Corp. *The Laquila Group Inc.		World Trade Center Recovery Project; 63rd Street Tunnel
Larkin & Sangster Lathrop, Shea & Henwood Company Le Roi Equipment Sales Company, Inc. Lee: Francis A. Lee Company, A Corporation		Con Ed East River Generating Station; FDR Drive Rehab; Citizen Gate Tunnel Repair (Keyspan gas main under Gowanus Canal)
Lehigh Portland Cement Co. Lenart Constructors, Inc. Leopold: J. Leopold & Co., Inc.		Paving Eastern Blvd and Whitlock Ave for original Triborough Bridge construction.

Members of the GCA of New York

(*Current GCA members)

(**Founding GCA members)

Original Company Name	Sample Projects
Lewis & McDowell, Inc.	
Ley: Fred T. Ley & Co., Inc.	Pennsylvania Station demolition; Third Avenue El demolition
Lidgerwood Manufacturing Company	
Linde & Griffith Co.	
Lipsett Division	
Litchfield Construction Company	
Lizza & Sons, Inc.	Roosevelt Island Bridge
Lockjoint Pipe Company	
Long's Construction & Dev. Co.	
Lopier Construction Corporation	
Lord Construction Company	North Shore Marine Transfer Station
Lucius Engineering Company	
Ludington: I. M. Ludington's Sons, Inc.	
Lynn Construction Co.	
Lyons-Slattery Co., Inc.	Ashokan Reservoir
M. & L. Contracting Co.	
M. C. Hauling Co.	
Mac Asphalt Construction Corp.	
MacArthur Brothers Company	
MacArthur Concrete Pile Corp.	Lincoln Tunnel; Water Tunnel No. 3
MacIsaac: F. J. MacIsaac	
Mack: B.T. & J. J. Mack, Inc.	
MacKay Construction Corp.	
MacLean; Grove & Company	
*Macro Enterprises LTD.	
Mahoney: James Mahoney, Inc.	
Malcolm: S. V. R. Malcolm & Son	
Malvese: George Malvese & Co., Inc.	Malvese Equipment Co Inc.
Manhattan Water Works Inc.	
Manitowoc-Forsythe Corporation	
Manning: Frank D. Manning	
Marcus Excavation & Foundation Corp.	Prospect Expressway
Marcus Substructure Corporation	Lincoln Tunnel; Brooklyn-Battery Tunnel; Catskills Aqueduct
Marcus: The Marcus Contracting Co., Inc.	
Mascali: Frank Mascali & Sons, Inc.	
Mason & Hanger Company	
Mason, Hilton & Company	
*Maspeth Supply Co., LLC.	
Massachusetts Bonding and Insurance Co.	LaGuardia Airport runway strengthening (2000)
Master Waterproofers, Inc.	Newtown Creek upgrade (2003)
Matac Construction Corporation	
*Max-Tec Construction Corp.	
*McCullagh: James McCullagh Co., Inc.	
*McCloskey Contracting Co., Inc.	
McCooey & Schmitz, Inc.	
McCormack: William J. McCormack Sand Co., Inc.	
McDonald: Charles S. McDonald	Wm. P. McDonald Construction Co.
McElroy & Kerwin, Inc.	Water Tunnel No. 2
McGovern: Patrick McGovern Company	

Members of the GCA of New York (*Current GCA members) (**Founding GCA members)	Original Company Name	Sample Projects
McInerny: M. R. McInerny Corp.		
McLean Contracting Company		
McMullen: Arthur McMullen		George Washington Bridge
Meads: Charles Meads & Co.		
Meehan Paving & Construction Co., Inc.	John Meehan & Son	Paving 77th Street to 112th Place for original Triborough Bridge construction; Holland Tunnel
Melrose Construction Co.		
Meltzer: Joseph Meltzer, Inc.		
Melwood Construction Corp.		
*Merco Inc./Mergentime		
Merrill-Ruckgaber Company		New York Central Harlem River Bridge
Merritt & Chapman Derrick & Wrecking Company	Merritt-Chapman & Scott Corporation	Throgs Neck Bridge; Verrazano-Narrows Bridge; Gansevoort Incinerator; New York Central Bridge over Harlem; Indian Point
*Metrotech Contracting Corp.		
Mideastern Contracting Corporation		
Miele: Joseph Miele Construction Co., Inc.		
Mill Basin Asphalt Corp.		
Miller: I. B. Miller Cont'g Corp.		
Milliman & Nazzaro Construction Corp.	Nazzaro Construction Corp.	Fountain Avenue Landfill remedial action
Modern Continental		
Mohawk Constructors, Inc.		
Monks: John Monks & Sons		
Moran and Fenneman		
Moran Bros. Contracting Co., Inc.		
Moran Towing & Transportation Company		Battery Park City ball fields (2003) (floating derrick)
Moranti & Faymond, Inc.	Paul J. Moranti, Inc.	
*Moretrench American Corp.		World Trade Center (dewatering of original bathtub and dewatering for reconstruction efforts); Springfield Gardens Industrial Park, Verrazano Bridge, Lenox Avenue Rehabilitation, Water Tunnel No. 3 Shaft Freezing
*Moriarty: T. Moriarty & Son, Inc.		Grand Central State rehabilitation; Waldorf Astoria rehabilitation
Morrison-Knudsen Company, Inc.		
Murphy: Charles F. Murphy, Jr.		
MWN Associates Inc.		
Myers & McWilliams		
N.Y. Submarine Contracting Co., inc.		
*NAB Construction Corp.		Narrows Water Tunnel; Water Tunnel No. 3; Greenpoint Avenue Bridge replacement
Nassau Construction Co., Inc.		
National Bank of North America		
National Excavation Corporation		North River Sewage Treatment Plant; Water Tunnel No. 3; Astoria Energy Power Plant; Con Ed East River Repowering Project; Brooklyn Navy Yard Cogeneration Facility; Astoria Generating Station; New York City Asphalt Plant; Flushing CSO Project
National Preferred Risks, Inc.		
National Structures Corporation		
National Surety Company		
Nazareth Cement Company		Whitestone Bridge; Belt Parkway

Members of the GCA of New York (*Current GCA members) (**Founding GCA members)	Original Company Name	Sample Projects
Oakhill Contracting Co., Inc.		
O'Brien Brothers, Inc.		
*Ocean Marine Development Corp.		
O'Connell: J. P. O'Connell & Co.		
O'Connor: L. P. O'Connor, Inc.		
O'Rourke Engineering Construction Company		
Ottaviano: A. E. Ottaviano, Inc.		
Overseas Machinery Suppliers Corp.		
*P.C.M. Contracting Co., Inc.		
Pacchiosi Drill USA Inc.		
Paino: Angelo Paino		
Paquet: M. J. Paquet Inc.		
Parisi: M. Parisi & Son, Inc.		
Park Contracting Corporation		
Patelli & Wilson		
Paterno & Sons, Inc.		
Paul Revere Life Insurance Co.		
Paulsen Construction Corp.		
Peerless Construction Company		
Pegno: A.J. Pegno Construction Corp.		
Pennsylvania Cement Co.	Pennsylvania-Dixie Cement Corporation	
*Perini Corporation	Perini & Sons, Inc.	Pennsylvania Station/tunnels
		63rd Street tunnel
		Hunts Point Sewage Treatment Plant
		Whitestone Expressway Bridge Replacement 2003
		Delaware Aqueduct; Narrows Water Tunnel; Newtown Creek Sewage Treatment Plant Upgrade 2003; AirTrain 2002; Water Tunnel No. 3
		LIRR locomotive shop (2001)
		World Trade Center Transportation Hub (2006)
*Petracca & Sons, Inc.		
Phoenix Construction Company	Phoenix Construction Associates	FDR Rehab; Coney Island Sewage Treatment Plant; Jerome Park Reservoir; Water Tunnel No. 3; LaGuardia Airport Runway Extensions; Owl's Head Sewage Treatment Plant renovation; Oak Point Link; DEP Water Pollution Control Plants (WPCP) including Newtown Creek, Tallman Island, Huntspoint, Wards Island
Phoenix Sand & Gravel Company		
Picone: John P. Picone Incorporated		
*Pile Foundation Construction Co., Inc.		Henry Hudson Parkway; Major Deegan Expressway; Newtown Creek Sewage Treatment Plant—and interceptor sewer under East River and Manhattan; North River Sewage Treatment Plant; West Side Highway/Miller Highway; Washington Bridge reconstruction (1949) Indian Point (site clearing, access roads)
Pinebrook Construction Co.		
Pleasantville Constructors, Inc.		
Pneumatic Tool Sales & Repair Co., Inc.		
Poirier & McLane Corporation		
*Posillico Civil, Inc.		
Post & McCord, Inc.		Brooklyn Army Terminal, Original Penn Station; Transcontinental Gas Pipeline's third tunnel (in Manhattan, 1958); West Side Interceptor Sewer
Powers-Kennedy Contracting Corp.		
Presscrete: The Presscrete Co.		
Preston Contracting Co., Inc.	Spearin, Preston & Burrows, Inc.	

Members of the GCA of New York

(*Current GCA members)
(**Founding GCA members)

Members of the GCA of New York	Original Company Name	Sample Projects
*Prima Paving Corp.		Central Park: The Great Lawn, Turtle Pond, Summit Rock, 72nd St. Playground, Greyshot Arch/Cedar Hill/ Glade Arch; Bakersfield Stadium Columbia University; Fordham University Murphy Field; MultiPhase Extraction for American Airlines at JFK Airport; Continental Airlines Cargo; HS/MS 368 Bronx; Port Authority Bus Terminal; New US Court House Brooklyn; Stuyvesant Town Improvements; Hudson River Park.
*Pristine Management Corp.		
Queens Structure Corp.		
Quinn-Meissner, Inc.		
Quist: Henry Quist, Inc.		
Ragonetti: John Ragonetti		
Rahmani Construction Corporation		
*Railroad: The Railroad Construction Family of Companies		
Rapid Transit Subway Construction Company		
Raylin Construction Corp.		
Raymond Concrete Pile Company		
*Reicon Group, LLC.	Reinauer Marine Construction Co.	Throgs Neck Bridge; Transcontinental Gas Pipeline; Howland Hook Express Rail Staten Island; Howland Hook Marine Terminal; LIRR Jamaica
Reid: Hugo Reid		
Reiss & Weinsier, Inc.	Peter Reiss Construction Co., Inc.	Rikers Island Bridge; North River Sewage Treatment Plant, Jacob Javits Convention Center Foundation; Northshore Marine Transfer Station Substructure Rehab; East 60th Street Helipad Substructure; Pier 16 SSP Ferry Landings; Brooklyn Port Authority Piers' Substructures; Sheepshead Bay Bulkhead Renewal
Reuss: Edward H. Reuss, Jr.		
Rhulen Agency, Inc.		
Rice & Ganey, Inc.		
Rice: Laurence J. Rice, Inc.	Rice Contracting Co., Inc.	Queens Approach piers for original Triborough Bridge, Coney Island Sewage Treatment Plant; intercepting sewers from northern Manhattan and south Bronx to Wards Island Erie Basin: North Shore Marine Transfer Station; Brooklyn Incinerator; Southwest Brooklyn Marine Transfer Station: East River heliport; North River heliport; Chelsea Piers; Battery Park City landfill; South Street Seaport piers
Richards & Gaston, Inc.		
Rivara: Anthony Rivara & Sons		
*River Pile & Foundation Co. Inc.		
Riverdale Contracting Co., Inc.		
Riverdale Water Works Corp.		
*Riverview Construction., Inc.		
*RMSK Contracting Corp.		
Road Material Corporation		
Roberts: E. O. Roberts Company, Inc.		
Rock Industries, Inc.		
Rockport Granite Company		
Rodgers & Hagerty, Inc.		
Rogers: George C. Rogers & Co., Inc.		
Rosenthal Engineering Contracting Co., Inc.		
Rosoff Sand & Gravel Corp.	Rosoff-Brader Construction Corporation	Coney Island Sewage Treatment Plant

Members of the GCA of New York (*Current GCA members) (**Founding GCA members)	Original Company Name	Sample Projects
Rosoff Subway Construction Co., Inc.	Samuel R. Rosoff, Ltd.	
Roth: Alfred P. Roth		
Rusciano & Son Corp.	Rusciano & Del Balso	Whitestone Bridge; Major Deegan Expressway; George Washington Bridge
Russgood Construction Co., Inc.		
*Ruttura & Sons Construction Co., Inc.		
*RWKS Transit Inc.		
Ryan Ready Mixed Concrete Corp.		
Ryan-Turecamo, Inc.	Vincent Turecamo, Inc.	JFK Airport Cargo area (2003) AirTrain (2001)
Savin: The Savin Construction Corporation		Delaware Aqueduct
*Schiavone Construction Co.		Water Tunnel No. 3; Croton Water Treatment Plant 2005; South Ferry Terminal Subway Station 2005; North River Sewage Treatment Plant, Atlantic Avenue Station rehab.
Schneider: H. T. Schneider, Inc.		
Scott Welded Products, Inc.		
*SDL Construction Corp.	Blandford Land Development Corp.	Delaware Aqueduct
Seaboard Construction Corporation		
Seaboard Sand & Gravel Corp.		
Seabrook: C. F. Seabrook Company, Inc.		
Senior and Palmer, Incorporated		New York Central Harlem River Bridge; George Washington Bridge; Wards Island Pedestrian Bridge; Unionport Bridge
Serber: D. C. Serber, Inc.		
*Shea: J.F. Shea Co., Inc.		Delaware Aqueduct; 7 Line extension
Sheehan: T. J. Sheehan Company, Inc.		
Shell Oil Company		
Sherwin: H. H. Sherwin & Co., Inc.		
Sicilian Asphalt Paving Company		
*Silverite Construction Co., Inc.		Queens Midtown tunnel rehabilitation (1996 and 2003), Newtown Creek WPCP (2008), Coney Island STP (1994, 1996, 2002); Rehabilitation Babylon Railroad Station (2003); 57th Street and Cauldwell Avenue Bridges (2005)
Sinnott & Canty, Inc.		
*Skanska Koch, Inc.	Karl Koch Erecting, Koch Skanska	Erection of World Trade Center, Upper Roadway Replacement of Manhattan Bridge (1959 and 2003), Newtown Creek WPCP (2008), JFK Light Rail System (AirTrain), Yankee Stadium (new stadium and 1970s rehab of old stadium), Erection of Javits Convention Center
*Skanska Mechanical and Structural Inc.	Gottlieb Mechanical	Water Tunnel No. 3; East River Repowering Plant; NYPA 500; Ravenswood
*Skanska USA Civil Northeast Inc.	Slattery Contracting Co., Inc.	Alexander Hamilton Bridge; Verrazano-Narrows Bridge: Goethals Bridge; Outerbridge Crossing; Coney Island Sewage Treatment Plant; Grand Central Parkway; Cross Bay Parkway; Kosciuszko Bridge; Gowanus Parkway; Bowery Bay Sewage Treatment Plant; Major Deegan Expressway; Hunts Point Sewage Treatment Plant; Brooklyn Queens Expressway; Alexander Hamilton Bridge; Verrazano-Narrows Bridge; North River Sewage Treatment Plant; Broadway Bridge; Northern Boulevard Bridge; Bruckner Boulevard Expressway; Long Island Expressway; Horace Harding Expressway; Clearview Expressway; Clove Lakes Expressway; West Shore Expressway; Harlem River Drive; JFK Light Rail; Highbridge Yard

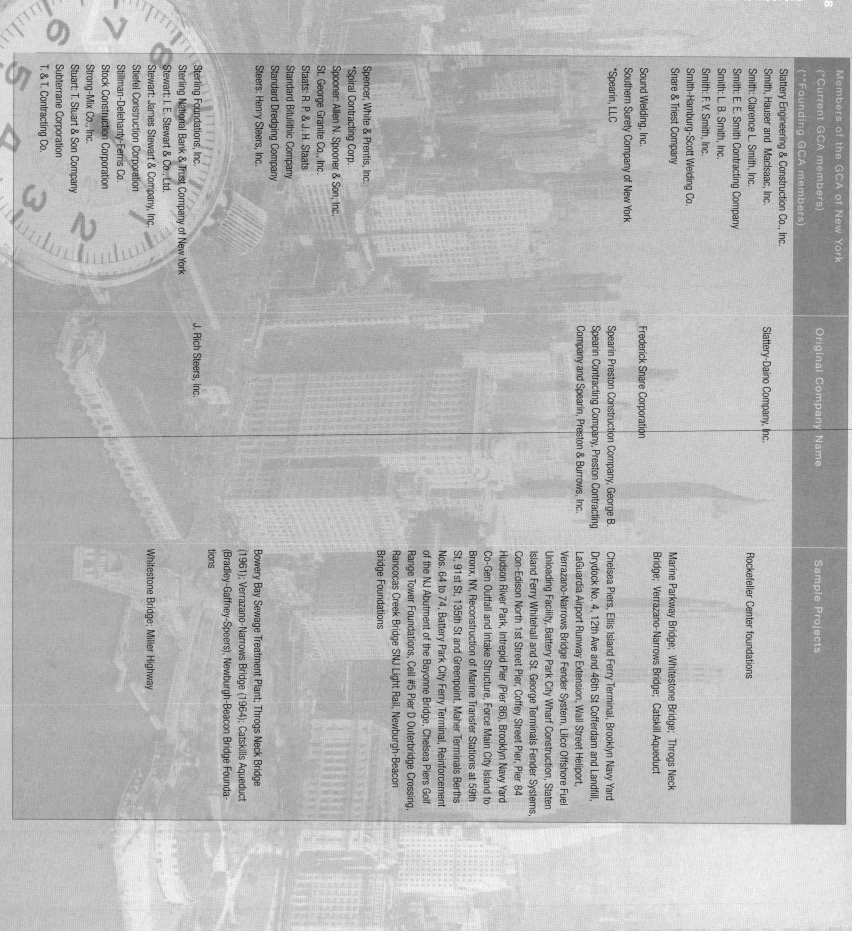

Members of the GCA of New York (*Current GCA members) (**Founding GCA members)	Original Company Name	Sample Projects
Slattery Engineering & Construction Co., Inc.	Slattery-Daino Company, Inc.	Rockefeller Center foundations
Smith, Hauser and MacIsaac, Inc.		
Smith: Clarence L. Smith, Inc.		
Smith: E. E. Smith Contracting Company		
Smith: L. B. Smith, Inc.		
Smith: F. V. Smith, Inc.		
Smith-Hamburg-Scott Welding Co.		
Snare & Triest Company		Marine Parkway Bridge; Whitestone Bridge; Throgs Neck Bridge; Verrazano-Narrows Bridge; Catskill Aqueduct
Sound Welding, Inc.		
Southern Surety Company of New York		
*Spearin, LLC	Spearin Preston Construction Company, George B. Spearin Contracting Company, Preston Contracting Company and Spearin, Preston & Burrows, Inc.	Chelsea Piers, Ellis Island Ferry Terminal, Brooklyn Navy Yard Drydock No. 4, 12th Ave and 46th St Cofferdam and Landfill, LaGuardia Airport Runway Extension, Wall Street Heliport, Verrazano-Narrows Bridge Fender System, Lilco Offshore Fuel Unloading Facility, Battery Park City Wharf Construction, Staten Island Ferry Whitehall and St. George Terminals Fender Systems, Con-Edison North 1st Street Pier, Coffey Street Pier, Pier 84 Hudson River Park, Intrepid Pier (Pier 86), Brooklyn Navy Yard Co-Gen Outfall and Intake Structure, Force Main City Island to Bronx, NY, Reconstruction of Marine Transfer Stations at 59th St, 91st St, 135th St and Greenpoint, Maher Terminals Berths Nos. 64 to 74, Battery Park City Ferry Terminal, Reinforcement of the NJ Abutment of the Bayonne Bridge, Chelsea Piers Golf Range Tower Foundations, Cell #5 Pier D Outerbridge Crossing, Rancocas Creek Bridge SNJ Light Rail, Newburgh-Beacon Bridge Foundations
Spencer, White & Prentis, Inc.	Frederick Snare Corporation	
*Spiral Contracting Corp.		
Spooner: Allen N. Spooner & Son, Inc.		
St. George Granite Co., Inc.		
Staats: R. P. & J. H. Staats		
Standard Bitulithic Company		
Standard Dredging Company		
Steers: Henry Steers, Inc.	J. Rich Steers, Inc.	Bowery Bay Sewage Treatment Plant; Throgs Neck Bridge (1961); Verrazano-Narrows Bridge (1964); Catskills Aqueduct (Bradley-Gaffney-Speers); Newburgh-Beacon Bridge Foundations
Sterling Foundations, Inc.		
Sterling National Bank & Trust Company of New York		
Stewart: I. E. Stewart & Co., Ltd.		
Stewart: James Stewart & Company, Inc.		
Stiefel Construction Corporation		
Stillman-Delehanty-Ferris Co.		
Stock Construction Corporation		
Strong-Mix Co., Inc.		Whitestone Bridge; Miller Highway
Stuart: T. Stuart & Son Company		
Subterrane Corporation		
T. & T. Contracting Co.		

Members of the GCA of New York
(*Current GCA members)
(**Founding GCA members)

Original Company Name	Sample Projects
Tadco Construction Corp.	
Tartarus Construction Company, Inc.	
Teller Paving & Contracting Corp.	
Terry & Tench Company, Inc.	
Terry Contracting, Incorporated	
*Thalle Industries, Inc.	
Thalle Construction Company	New York Central Harlem River Bridge; Prospect Expressway Amawawalk and Titcus Dams reconstruction (1998), Bowery Bay Sewage treatment plant, Boyds Corner Dam, The Bronx Zoo Baboon Exhibit, Ashokan Reservoir and Roundout Reservoir Pavement Reconstruction, Katan Avenue Interceptor Sewer, Kensico Watershed storm water management
*The Railroad Construction Family of Companies	
The Robbins-Ripley Co.	
The Tuller Construction Company	
The Utah Construction Company, Inc.	
Thompson-Leopold-Fredburn Engineering Co., Inc.	
Timoney: J. V. Timoney Construction Co.	
Tomasetti: Ralph Tomasetti and Co.	
Tomkins: Calvin Tomkins	
Tomkins Cove Stone Company	Delaware Aqueduct
Torpey: Michael J. Torpey, Inc.	
*Transit Construction Corp.	
Transit-Mix Concrete Corporation	
*Trevcon Construction Co., Inc.	
*Trevicos Corporation	Ferry Terminal at West 38th St. (2005); Brooklyn Cruise Terminal (2005), Manhattan Cruise Terminal substructure rehabilitation, Pier 25 (2007)
	Reconstruction of Brooklyn-Queens Expressway; Williamsburg Bridge Reconstruction; East Side Access: WTC East bathtub Slurry Wall, Paerdegat Basin
Tri-Boro Asphalt Corporation	
Triest Contracting Corporation	
Tufano Contracting Corporation	Queens Midtown Tunnel
*Tulger Contracting Corp.	
*Tully Construction Co., Inc.	
Tully & Di Napoli, Inc.	Bronx Kills and Bronx approach piers, excavation fill and demolition on Randall's Island, Bronx Approach ramp and paving, paving Wards and Randall's Islands Viaducts, for original Triborough Bridge Construction. 1939 World's Fair. Construction of original Con Ed plants on East River. 9/11 Cleanup: debris from N, R subway stations; Hudson River park; Tully-Pegno-Yonkers tri-venture built path at ground zero; Whitestone Bridge; Major Deegan Expressway; Van Wyck Expressway; Newtown Creek Sewage Treatment Plant; George Washington Bridge (Martha); Triborough Bridge; Grand Central Parkway, Cross Bronx Expressway, Southern State Parkway, Long Island Expressway, Idlewild/Kennedy Airport, LaGuardia Airport, foundation for the World Trade Center; West Side Highway; Whitestone Expressway Bridge Replacement 2003; Staten Island Transfer Station (2002); Columbus Circle Reconstruction (2004); 9A reconstruction in lower Manhattan Cross Island Parkway, Water Tunnel No. 3

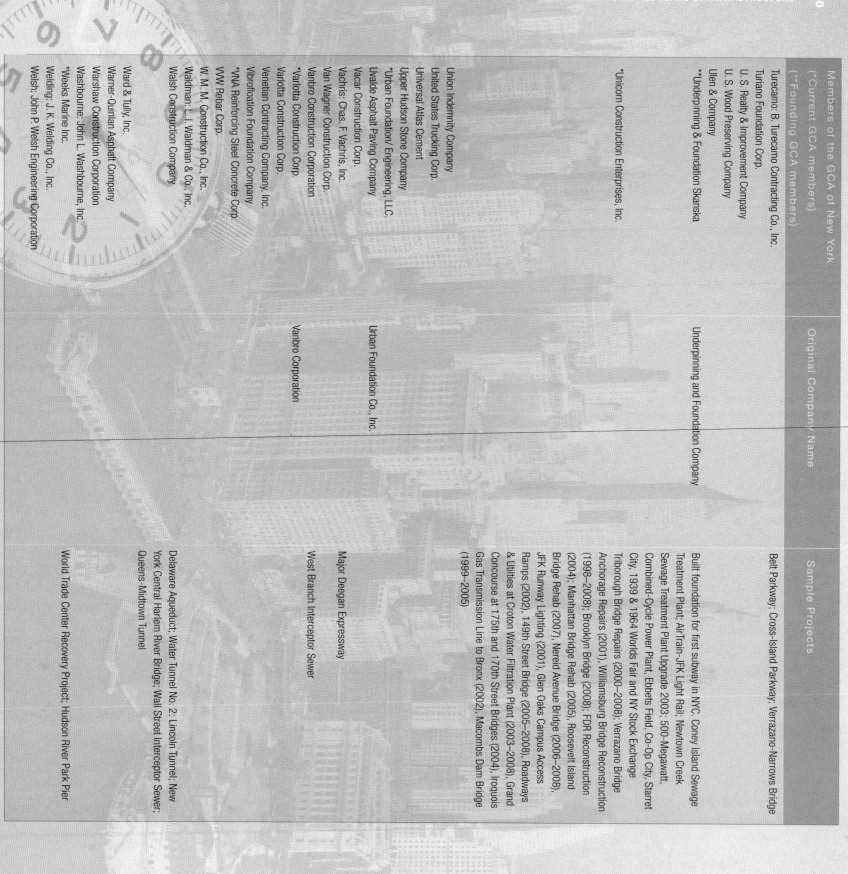

Members of the GCA of New York (*Current GCA members) (**Founding GCA members)	Original Company Name	Sample Projects
Turecamo: B. Turecamo Contracting Co., Inc.		Belt Parkway; Cross-Island Parkway; Verrazano-Narrows Bridge
Turiano Foundation Corp.		
U.S. Realty & Improvement Company		
U.S. Wood Preserving Company		
Ulen & Company		
**Underpinning & Foundation Skanska	Underpinning and Foundation Company	Built foundation for first subway in NYC; Coney Island Sewage Treatment Plant; AirTrain-JFK Light Rail; Newtown Creek Sewage Treatment Plant Upgrade 2003; 500-Megawatt, Combined-Cycle Power Plant, Ebbets Field, Co-Op City, Starret City, 1939 & 1964 Worlds Fair and NY Stock Exchange
*Unicorn Construction Enterprises, Inc.		Triborough Bridge Repairs (2000–2008); Verrazano Bridge Anchorage Repairs (2001), Williamsburg Bridge Reconstruction (1998–2008); Brooklyn Bridge (2008); FDR Reconstruction (2004); Manhattan Bridge Rehab (2005); Roosevelt Island Bridge Rehab (2007); Nereid Avenue Bridge (2006–2008), JFK Runway Lighting (2001), Glen Oaks Campus Access Ramps (2002), 149th Street Bridge (2005–2008), Roadways & Utilities at Croton Water Filtration Plant (2003–2008), Grand Concourse at 175th and 170th Street Bridges (2004), Iroquois Gas Transmission Line to Bronx (2002), Macombs Dam Bridge (1999–2005)
Union Indemnity Company		
United States Trucking Corp.		
Universal Atlas Cement		
Upper Hudson Stone Company		
*Urban Foundation/Engineering, LLC.	Urban Foundation Co., Inc.	Major Deegan Expressway
Uvalde Asphalt Paving Company		
Vacar Construction Corp.		
Vachris: Chas. F. Vachris, Inc.		
Van Wagner Construction Corp.		
Vanbro Construction Corporation	Vanbro Corporation	West Branch Interceptor Sewer
*Variotta Construction Corp.		
Variotta Construction Corp.		
Venetian Contracting Company, Inc.		
Vibroflotation Foundation Company		
*VNA Reinforcing Steel Concrete Corp.		
VW Rebar Corp.		
W. M. M. Construction Co., Inc.		
Waldman: L. I. Waldman & Co., Inc.		
Walsh Construction Company		Delaware Aqueduct: Water Tunnel No. 2; Lincoln Tunnel; New York Central Harlem River Bridge; Wall Street Interceptor Sewer; Queens-Midtown Tunnel
Ward & Tully, Inc.		
Warner-Quinlan Asphalt Company		
Warshaw Construction Corporation		
Washbourne Corporation		
Washbourne: John L. Washbourne, Inc.		
*Weeks Marine Inc.		World Trade Center Recovery Project; Hudson River Park Pier
Welding: J. K. Welding Co., Inc.		
Welsh: John P. Welsh Engineering Corporation		

Members of the GCA of New York

(*Current GCA members)

(**Founding GCA members)

Members of the GCA of New York	Original Company Name	Sample Projects
West: Sid. J. West		
Western Concrete Pile Corporation/Western Foundation Company		
Weston: Andrew Weston Co., Inc.		
Wetesen: Thor. Wetlesen		
Wheeler: James Wheeler, Inc.		
Wheeling Mold & Foundry Company		
White: J. P. White Company		
White: The White Motor Company		
Wilcox & Stutz, Inc.		
*William A. Gross Construction Associates		
Wilson & English Construction Company		
Wirston & Company		
Wirston Bros. Company		
Wood Powder Company, Inc.		
Wood: R. D. Wood & Company		
Woodcrest Construction Co., Inc.		
Worthington	Western Contracting Company	Ashokan Reservoir/Croton Falls Reservoir
Yecmans-Drews Corporation		
*Yonkers Contracting Co., Inc.		Red Hook Sewage Treatment Plant; New England Thruway (Bronx section opened 1958); Greenpoi't Avenue Bridge (replacement, 1984)
Zaaa Contracting Co., Inc.		
Zeke Construction Corp.		
Zoel/Zena Company		

GCA Presidents

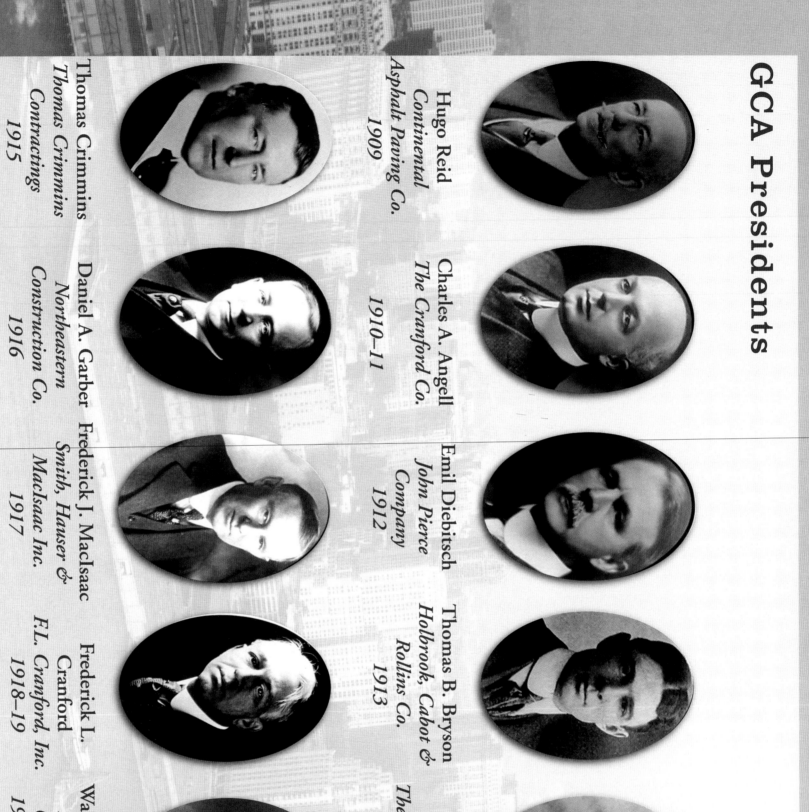

Hugo Reid
Continental
Asphalt Paving Co.
1909

Charles A. Angell
The Cranford Co.
1910–11

Emil Diebitsch
John Pierce
Company
1912

Thomas B. Bryson
Holbrook, Cabot &
Rollins Co.
1913

Franklin
Remington
The Foundation Co.
1914

Thomas Crimmins
Thomas Crimmins
Contractings
1915

Daniel A. Garber
Northeastern
Construction Co.
1916

Frederick J. MacIsaac
Smith, Hauser &
MacIsaac Inc.
1917

Frederick L.
Cranford
F.L. Cranford, Inc.
1918–19

Walter J. Drummond
Beaver Engr. &
Contracting Co.
1920–21 and 1933

Arthur A. Johnson
*The Arthur A.
Johnson Corp.
1928–29*

William V. McMenimen
*Raymond Concrete Pile
Co.
1926–27*

Walter H. Gahagan
*W.H. Gahagan, Inc.
1925*

Patrick McGovern
*Patrick McGovern,
Inc.
1924*

John J. Watts
*Mason & Hanger
Co.
1923 and 1933*

James L. Carey
*Necaro Company
Inc.
1935*

Ray N. Spooner
*Allen N. Spooner &
Son, Inc.
1934*

E.A. Herrick
*Oakdale
Contracting Co.
1932*

C. Aubrey Nicklas
*Empire
Construction Co.
1931*

Harry Haggarty
*The Sicilian
Asphalt Paving Co.
1930*

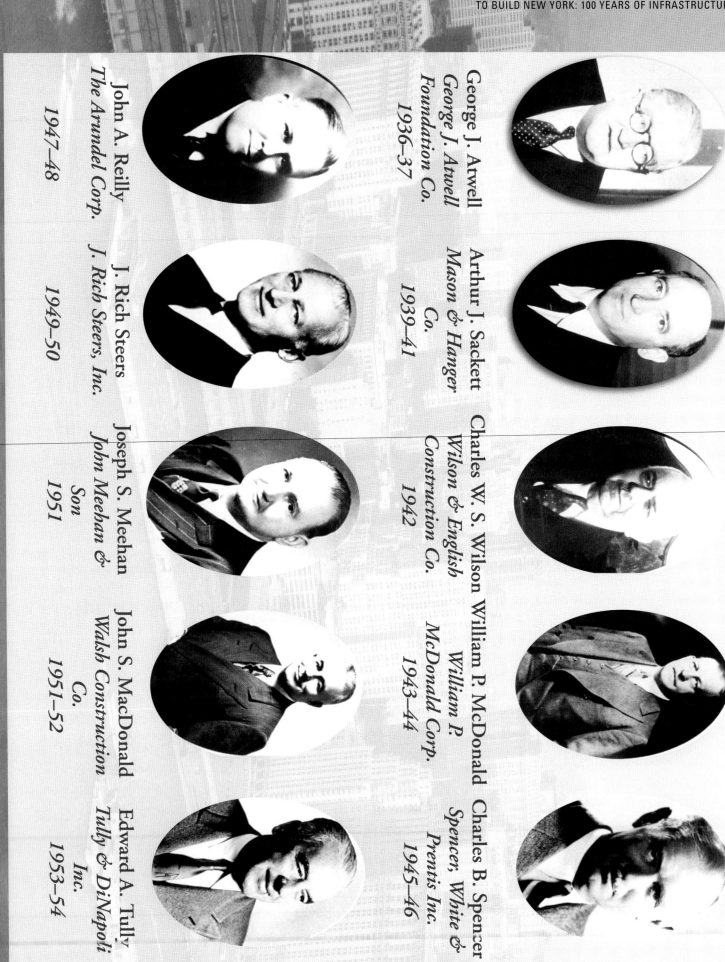

George J. Atwell
George J. Atwell
Foundation Co.
1936–37

Arthur J. Sackett
Mason & Hanger
Co.
1939–41

Charles W. S. Wilson
Wilson & English
Construction Co.
1942

William P. McDonald
William P.
McDonald Corp.
1943–44

Charles B. Spencer
Spencer, White &
Prentis Inc.
1945–46

John A. Reilly
The Arundel Corp.
1947–48

J. Rich Steers
J. Rich Steers, Inc.
1949–50

Joseph S. Meehan
John Meehan &
Son
1951

John S. MacDonald
Walsh Construction
Co.
1951–52

Edward A. Tully
Tully & DiNapoli
Inc.
1953–54

Joseph J. Haggerty
The Sicilian
Asphalt Paving Co.
1955–56

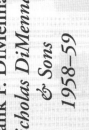

Richard E. Talmadge
MacArthur
Concrete Pile Co.
1957

Frank P. DiMenna
Nicholas DiMenna
& Sons
1958–59

Mansell L. MacLean
MacLean Grove &
Co., Inc.
1960–61

Howard A. Collins
H & P Excavation &
Foundation Co., Inc.
1962

Edward A. White
Spencer, White
Prentis, Inc.
1970–71

Robert P. Bayard
Johnson, Drake &
Piper Inc.
1963

George W. Beinsch
Slattery
Contracting Co., Inc.
1964–65

Anthony G. Gull
Gull Contracting
Co, Inc.
1966–67

Moses Hornstein
Horn Construction
Co., Inc.
1968–69

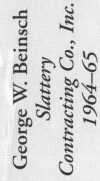

Daniel M. Lazar
*Cayuga
Construction Co.*
1972–73

John D. Saunders
*Slattery Associates,
Inc.*
1974–75, 1988

Gerard P. Tully
*Willets Point
Contracting Corp.*
1976–77

Gerard Neumann
*Spearin, Preston &
Burrows Inc.*
1978–79

George A. Fox
*Grow Tunneling
Corp.*
1980–81

Nicholas DiMenna
*Nicholas DiMenna
& Sons, Inc.*
1982–83

Theodore Civetta
*John Civetta &
Sons, Inc.*
1984–85

William Goodman
*Grow Tunneling
Corp.*
1986–87

Kenneth Tully
*Tully Construction
Co.*
1989–90

Gerard Neumann Jr.
*Spearin, Preston &
Burrows Inc.*
1991–92

Peter Tully
*Tully Construction
Co., Inc.*
2001–2002

Thomas King
*Schiavone
Construction Co., Inc.*
1999–2000

Edward Simpson
*Nab Construction
Corp.*
1997–98

James Moriarty Jr.
*T. Moriarty & Sons,
Inc.*
2007–2008

Donald Unbekant
Perini Corporation

1995–96

Thomas Iovino
*Judlau
Contracting, Inc.*
2004–2006

Edward Cruz
*EE Cruz & Co.,
Inc.*
1993–94

John Pegno
AJ Pegno & Sons

2003–2004

Photo Credits

Subject Matter	Page Number	Source
World Trade Center	Book Spine, Top	Skanska USA Civil Northeast
Grand Central Terminal	Book Spine, Middle	Getty Images
Verrazano-Narrows Bridge	Book Spine, Bottom	Skanska USA Civil Northeast
	Front Cover, Top Row (From Left to Right)	
New York City Water Tunnel No. 3		New York City Department of Environmental Protection
New York Power Authority 500MW Combined-Cycle Power Plant		Skanska USA Civil Northeast
Grand Central Station		Getty Images
Alexander Hamilton Bridge		Skanska USA Civil Northeast
Coney Island Wastewater Treatment Plant		Silverite Construction Co., Inc.
AirTrain		Skanska USA Civil Northeast
AirTrain		Skanska USA Civil Northeast
New York Power Authority 500MW Combined-Cycle Power Plant		Skanska USA Civil Northeast
World Trade Center	Front Cover, Middle Row (From Left to Right)	Skanska Koch
Verrazano-Narrows Bridge		Skanska USA Civil Northeast
Alexander Hamilton Bridge		Skanska USA Civil Northeast
Gowanus Expressway		Skanska USA Civil Northeast
Bruckner Interchange		Skanska USA Civil Northeast
Yankee Stadium	Front Cover, Bottom Row (From Left to Right)	Delaney Associates, LP
Seagrams Building		Port Authority of New York and New Jersey
First Regional Plan		New York Transit Museum
First Regional Plan		Regional Plan Association
Bruckner Interchange		Skanska USA Civil Northeast
IRT Brooklyn Line		Port Authority of New York and New Jersey
George Washington Bridge		United States Library of Congress Prints and Photographs Division
Force Main Sewer at JFK Airport		Skanska Koch
74th Street/Roosevelt Avenue Subway Station		Regional Plan Association
World Trade Center	Back Cover	Skanska Koch
First Regional Plan	6	MTA Bridges and Tunnels Special Archive
George Washington Bridge	6	Slattery Collection
Goethals Bridge (both photos)	6	New York City Department of Parks
Manhattan Bridge	6	MTA Bridges and Tunnels Special Archive
Queensboro Bridge (left)	6	MTA Bridges and Tunnels Special Archive
Queensboro Bridge (right)	6	MTA Bridges and Tunnels Special Archive
Triborough Bridge	6	Slattery
Alexander Hamilton Bridge	7	New York City Municipal Archives
Bronx-Whitestone Bridge	7	New York City Municipal Archives
New York Central Harlem Bridge	7	MTA Bridges and Tunnels Special Archive
Throgs Neck Bridge (both photos)	7	New York City Department of Environmental Protection
Verrazano-Narrows Bridge (left)	7	New York City Department of Environmental Protection
Verrazano-Narrows Bridge (center and right)	7	Silverite Construction Co.
26th Ward Wastewater Treatment Plant	8	New York City Municipal Archives
Ashokan Reservoir	8	New York City Municipal Archives
Bowery Bay Wastewater Treatment Plant (both photos)	8	New York City Department of Environmental Protection
Coney Island Wastewater Treatment Plant (both photos)	8	New York City Department of Environmental Protection
Hunt's Point Wastewater Treatment Plant	8	New York City Municipal Archives
Rondout Reservoir	8	New York City Municipal Archives
Ward's Island Wastewater Treatment Plant	9	New York City Department of Environmental Protection
Croton Water Filtration Plant	9	Schiavone Construction Corp.
DEP 6-inch Water Mains	9	New York City Department of Environmental Protection

158

Subject Matter	Page Number	Source
Water Tunnel Stage 1	9	New York City Department of Environmental Protection
Water Tunnel #3 (all photos)	9	New York City Department of Environmental Protection
Centre Street Loop	10	New York Transit Museum
Flushing Line (all photos)	10	New York Transit Museum
IRT Construction (both photos)	10	New York Transit Museum
63rd Street East River Tunnel (both photos)	11	New York Transit Museum
7 Train Extension	11	JF Shea
IND Eighth Avenue Line	11	New York Transit Museum
Second Avenue Tunnel	11	New York Transit Museum
Belt Parkway	12	New York City Department of Parks
Brooklyn-Queens Expressway	12	New York City Department of Parks
Gowanus Parkway	12	MTA Bridges and Tunnels Special Archive
Henry Hudson Parkway	12	Municipal Archives
Holland Tunnel (both photos)	12	Port Authority of New York and New Jersey
Hutchinson River Parkway	12	MTA Bridges and Tunnels Special Archive
Lincoln Tunnel (both photos)	12	Port Authority of New York and New Jersey
Queens Midtown Tunnel (both photos)	12	MTA Bridges and Tunnels Special Archive
Brooklyn Battery Tunnel	13	MTA Bridges and Tunnels Special Archive
Cross Bronx Expressway	13	MTA Bridges and Tunnels Special Archive
East River Drive	13	Municipal Archives
Van Wyck Expressway	13	MTA Bridges and Tunnels Special Archive

Notes

1909–1929

1 Ric Burns and James Sanders, New York: An Illustrated History. New York: Alfred A. Knopf, 1999, pp. 271-272.

2 See New York City Department of City Planning website, http://www.nyc.gov/html/dcp/html/zone/zonehis.shtml.

3 George H. Johnson, "Rapid Transit in Great Cities," *Frank Leslie's Popular Monthly*, July, 1900, p. 2.

4 *The New York Times*, April 17, 1900, p. 5.

5 Burns and Sanders *op cit*, p. 255.

6 *Wall Street Journal*, February 14, 1912, p. 3.

7 *The New York Times*, April 28, 1912, p. 14.

8 Charles Merguerian, "A History of the NYC Water Supply System," http://www.dukelabs.com/Abstracts%20and%20Papers/CM2000c.htm.

9 E.g: GCA Bulletin, 1-1951, p. 7; 2-1960, p. 8; NYT, 9-6-1913, p. 3; Lazarus C.E. White, The Catskill Water Supply of New York City New York: John Wiley and Sons, 1913, pp. 212, 386; Diane Galusha, Liquid Assets; A History of New York's Water System, NY: Purple Mountain Press Ltd, 1999: ppp. 106, 107, 110, 115, 125ff, 131, 141, 154, 157, 176; Charles H. Wedner, Water for a City, New Brunswick, NJ: Rutgers UP, 1974, p.193.

10 http://www.tesoc.org/pubs/history2022/nov12.htm.

11 http://www.tesoc.org/pubs/history/2022/nov12.htm.

12 *The New York Times*, November 15, 1928, p.24.

13 NY and NJ Bridge and Tunnel Commission, 1931 [untitled pamphlet for opening day], p. 15.

14 E.g: GCA Bulletin, 1931, p. 219; NYT, 7-23-1954, p. 23.

15 Slattery Associates, Inc., Slattery: Rebuilding America's Infrastructure, nd; GCA Bulletin, 1927, p. 180.

16 E.g: Pennsylvania Railroad, The New York Improvement and Tunnel Extension of the Pennsylvania Railroad, Philadelphia: PRR, 8-1910, p. 22; Benjamin Miller, Fat of the Land: Garbage in New York, the Last Two Hundred Years, NY: Four Walls Eight Windows, 2000.

17 NYT, 9-14-1975, p. 45.

18 Flagg, Thomas R., Brooklyn Army Supply Base, Upper New York Bay from Fifty-eighth to Sixty-fourth Streets, Brooklyn, Kings County, NY 9-1988, http://hdl.loc.gov/loc.pnp/hhh.ny1594.

19 *The New York Times*, June 2, 1914, p. 14.

20 NYT, 3-6-1920, p. 15.

21 *Chicago Daily Tribune*, December 15, 1917, p. 3.

22 *Washington Post*, November 25, 1917, p. 1.

1930–1945

1 See Robert Caro, The Power Broker: Robert Moses and the Fall of New York," New York: Alfred A. Knopf, 1974.

2 http://www.bullard.com/company/hardhathistory.shtml.

3 GCA Bulletin, 1937, pp. 59, 154; GCA Bulletin, 3-1964, p. 12; Slattery Associates, Inc., "Slattery: Rebuilding America's Infrastructure, nd.

4 http://www.nycroads.com/crossings/triborough.

5 GCA Bulletin, 1936, pp. 67, 243; 3-1964, p. 12.

6 GCA Bulletin, 1939, p. 57; 1-1964, p. 24.

7 GCA Bulletin, 1929, p. 109; 1930, p. 183; 1933, p. 17.

8 GCA Bulletin, 1937, p. 155; 1941, p. 131; 1953, p. 4; 1954, p. 13.

9 http://www.panynj.gov/AboutthePortAuthority/HistoryofthePortAuthority/.

10 Kevin Baker, "Ideas & Trends: Recycling in New York; The History of Ash Heaps," *The New York Times*, January 5, 2003.

11 John W. Harrington, *The New York Times*, August 25, 1935, p. E10.

1946–1962

1 http://www.timessquarenyc.org/about_us/events_vjday.html.

2 Ric Burns and James Sanders, with Lisa Ades, New York: An Illustrated History. New York: Alfred A. Knopf, 1999.

3 GCA Bulletin, 3-1949, p. 9.

13 GCA Bulletin, 1942. P. 77.

14 GCA Bulletin, 1943, p. 74.

4 Robert Caro, The Power Broker: Robert Moses and the Fall of New York. New York, Vintage Books, 1975, p. 927.
5 In 1947–1948, 3.3 billion riders a year rode the city's subways and buses; by 1977, only 1.6 billion did. Frank J. Prial, "High Court Didn't Help; Transit Planning Is Near A Standstill," *The New York Times*, May 8, 1977, p. 134.
6 Robert Moses, "The Traffic Menace, in both Peace and War," *The New York Times*, April 29, 1951, p. SM8.
7 Sam Schwarts and Roy Cottam coined the term in 1980. (http:www.gridlocksam.com/about/html).
8 Sam Falk, "Tube to Brooklyn Will Open Today; New Brooklyn–Battery Tunnel Ready to Handle Today's Opening Traffic," *The New York Times*, May 25, 1950.
9 GCA Bulletin, 9-1949, p. 13; NYT, 8-29-1958; Slattery Associates, op. cit.
10 Merrill Folsom, NYT, 77-14-1962, p. 30.
11 GCA Bulletin, 11–1957, p. 21; 3–1961, p. 15.
12 Robert Moses (citing Corps of Engineers data), *The New York Times*, March 7, 1948, p. SM16.
13 Charles G. Bennett, "Rains Help Little as Savings in City Aid Water Supply," *The New York Times*, December 15, 1949, p. 1.
14 Thomas Buckley, "Low Bid for Hudson Water Station Surprises City," *The New York Times*, September 11, 1965, p. 28.

Underpinning
1 Robert Ridgway, Chief Engineer, Board of Transportation, Engineering New York's New Subway, GCA Bulletin, 1925, pp. 188-191.

1963–1979
1 http://www.nytimes.com/specials/nyc100/nyc100-8-haberman.html.
2 http://en.wikipedia.org/wiki/New York City Blackout of 1977.
3 Thomas M. Brosnan and Marie L. O'Shea, "Long–Term Improvements in Water Quality Due to Sewage Abatament in the Lower Hudson River," *Estuaries*, Vol. 19, No. 4 (December, 1996), p. 892.
4 "City Subway Gets $99 Million Grant," *The New York Times*, October 26, 1969, p. 46.
5 Videotaped interview with Felice Farber of GCA and Nicole Korkolis of the Carmen Group.

Concrete
1 David Owen, "Concrete Jungle," *New Yorker*, November 10, 2003, pp. 62ff; www.usbr.gov/LC/region/pao/brochures/hoover.html.
2 Vaclav Smil, *Creating the Twentieth Century: Technical Innovations of 1867–1914 and Their Lasting Impact*; Cary, NC: Oxford University Press, 2005, p. 311.
3 *Scientific American*, May 12, 1906, pp. 388-389.

1980–2009
1 Jack Rosenthal, "The Lives They Lived: Roger Starr, B. 1918, the Contrarian," *The New York Times Magazine*, December 30, 2001.
2 "Take Better Care of Less, New York City," *The New York Times*, April 18, 1979, p. A22.
3 James T. Wooten, "Aging Process Catches up with Cities of the North," *The New York Times*, February 13, 1976, p. 1.
4 "Take Better Care of Less, New York City," *The New York Times*, April 18, 1979, p. A22.
5 Pam Belluck, "New York City Has a) Too Many Pupils, b) Empty Seats, c) Both," *The New York Times*, September 8, 1996, p. 41.
6 "Housing Policy in New York City: A Brief History," Furman Center for Real Estate and Public Policy, New York University, 2006.
7 Richard Levine, "Economists Fault Population Figure for New York City," *The New York Times*, September 1, 1990, p. 1.
8 http://www.nyc.gov/html/nypd/pdf/chfdept/cscity.pdf.
9 Jamie Schram, *New York Post*, December 19, 2006.
10 Edward A. Gargan, "Rebuilding Plan for City Offered By Koch," *The New York Times*, May 13, 1981, p. B2.
11 Felicia R. Lee, "Dinkins Announces Reductions of $2.3 Billion in Capital Budget," *The New York Times*, October 13, 1990, p. 1.
12 Independent Budget Office, "Analysis of the Mayor's Executive Budget Capital Plan for 2001-2004," June, 2006.
13 William Neuman, "Long Planned, Transit Projects Get U.S. Help," The New York Times, December 19, 2006.
14 Ibid.
15 Videotaped interview with Felice Farber of GCA and Nicole Korkolis of the Carman Group, 2008.
16 Benjamin Miller, "Sanitation and Sewage," The Encyclopedia of New York State. Syracuse: Syracuse University Press, 2005.
17 Amy Waldman, "Despite Ruling, E.P.A. Insists that City Must Filter Water," *The New York Times*, February 10, 2001, p. B3; http.nyc.gov/html/news/croton.html.
18 http://www.buildings.com/Articles/detail.asp?ArticleD=31, p. 23.
19 www.msnbc.msn.com/id/3068158/site/newsweek.

Notes continued

1980–2009 (continued)

20 www.comptroller.nyc.gov/bureaus/bud/reports/impact-9-11-year-later.pdf.

21 Tom Frederickson, *Waste News*, February 18, 2002; http://graceindustriesinc.com/crisis.html.

23 The Richter-scale reading for the South Tower was 2.1. http://janedoe0911.tripod.com/StarWarsBeam2.html.

24 Perhaps the most serious damage was the breakage of nearly six miles of water mains, along with the severing of 300,000 telephone lires. Greg Gittrich, *New York Daily News*, December 11, 2001, p. 42.

25 Videotaped interview with Felice Farber of GCA and Nicole Korkolis of the Carmen Group, 2008.

26 Videotaped interview with Felice Farber of GCA and Nicole Kortkolis of the Carmen Group, 2008.

27 Videotaped interview with Felice Farber of GCA and Nicole Korkolis of the Carmen Group, 2008.

28 Videotaped interview with Felice Farber of GCA and Nicole Korkolis of the Carmen Group, 2008.

General Contractors Association

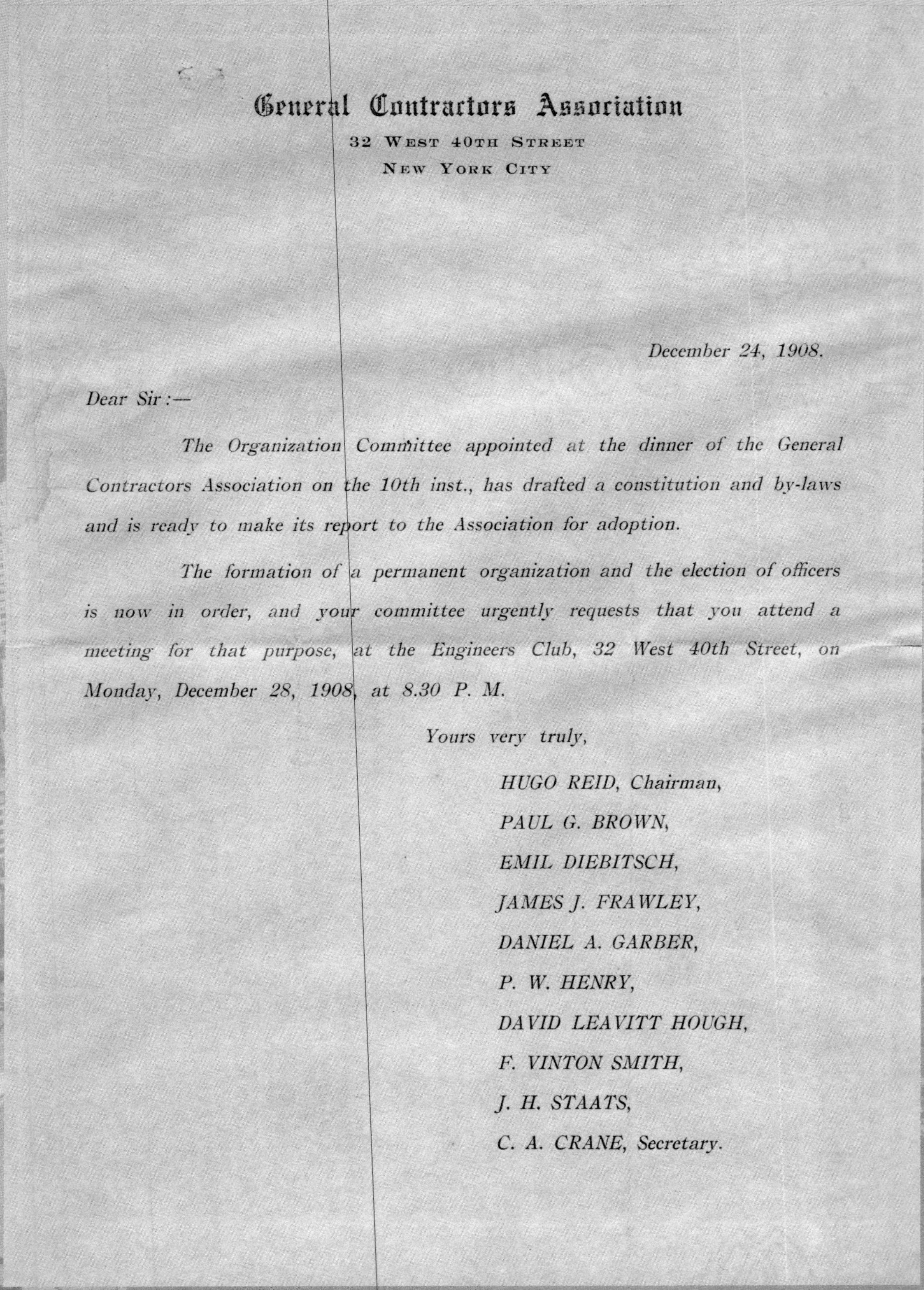

32 West 40th Street

New York City

December 24, 1908.

Dear Sir:—

The Organization Committee appointed at the dinner of the General Contractors Association on the 10th inst., has drafted a constitution and by-laws and is ready to make its report to the Association for adoption.

The formation of a permanent organization and the election of officers is now in order, and your committee urgently requests that you attend a meeting for that purpose, at the Engineers Club, 32 West 40th Street, on Monday, December 28, 1908, at 8.30 P.M.

Yours very truly,

HUGO REID, Chairman,

PAUL G. BROWN,

EMIL DIEBITSCH,

JAMES J. FRAWLEY,

DANIEL A. GARBER,

P. W. HENRY,

DAVID LEAVITT HOUGH,

F. VINTON SMITH,

J. H. STAATS,

C. A. CRANE, Secretary.

GCA Formation Leter (left)

December 24, 1908.

Courtesy General Contractors Association of New York.

154

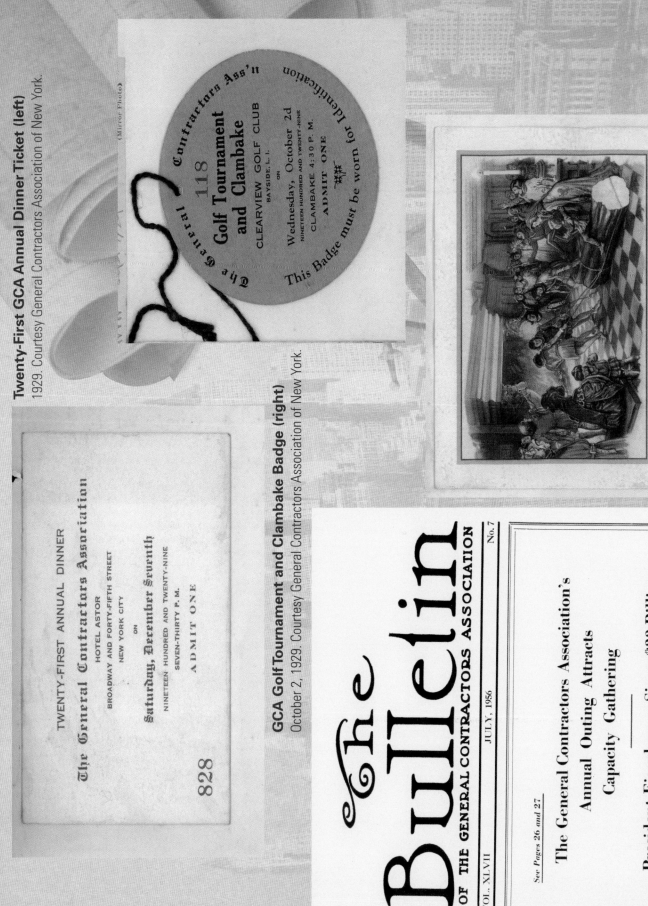

Twenty-First GCA Annual Dinner Ticket (left)
1929. Courtesy General Contractors Association of New York.

(Mirror Photo)

118

The General Contractors Ass'n

Golf Tournament and Clambake

CLEARVIEW GOLF CLUB
BAYSIDE, L. I.

ON

Wednesday, October 2d
NINETEEN HUNDRED AND TWENTY-NINE

CLAMBAKE 4:30 P. M.

ADMIT ONE

##

This Badge must be worn for Identification

TWENTY-FIRST ANNUAL DINNER

The General Contractors Association

HOTEL ASTOR
BROADWAY AND FORTY-FIFTH STREET
NEW YORK CITY

ON

Saturday, December Seventh
NINETEEN HUNDRED AND TWENTY-NINE

SEVEN-THIRTY P. M.

ADMIT ONE

828

GCA Golf Tournament and Clambake Badge (right)
October 2, 1929. Courtesy General Contractors Association of New York.

Christmas Greetings

GCA Christmas Card (above)
December 1928. Courtesy General Contractors Association of New York.

GCA Bulletin (left)
President Eisenhower signs Federal-Aid Highway Act. July 1956. Courtesy General Contractors Association of New York.

The Bulletin

OF THE GENERAL CONTRACTORS ASSOCIATION

VOL. XLVII   JULY, 1956   No. 7

See Pages 26 and 27

The General Contractors Association's
Annual Outing Attracts
Capacity Gathering

President Eisenhower Signs $33-Billion
Federal-Aid Highway Act

Other Features in this Issue

Looking Around

Activities of Our Members

Trade Notes and New Literature

Tax Law Notes —— Labor Law Notes

Alphabetical Index to Advertisers, Page 62   Classified Directory, Page 63

Index

A

A. J. Pegno Construction Corp., 125, 126
Abyssinian tubes, 68
Aimi, Akira, 95
Air quality, 98
Airlocks, 37, 39
AirTrain, 114, 121
Alexander Hamilton Bridge, 7, 79
AMEC, 125
American Pipe & Construction Company, 23
Ammann, Othmar, 25
Andrew Catapano Co., Inc., 54, 101, 123
Angel, Charles A., 152
Anthony Grace & Sons, Inc., 79
Aqueducts, 23, 24, 40
Argent Enterprises Corp., 123
Arthur A. Johnson Corporation, 54, 57, 79
Arthur McMullen Co., 25, 26, 54
Ashokan Reservoir, 8, 23
Asphalt Green park, 133
Asphalt paving, 130–133
Asphalt plants, 132, 133
Astoria natural gas plant, 29
Atwell, George J., 154

B

Battery Park City, 40, 105
Battery Park Seawall, 74, 75
Bayard, Robert P., 155
Beame, Abraham, 101, 105
Beaver Concrete, 126
Bedrock, 34–36, 44
Beinsch, George W., 155
Bell Telephone Company, 29
Belmont, August, 20
Belt Parkway, 12, 57
"Bends," 37, 39
Blackout (1977), 100
Blasting, 36, 42, 47
Blaw-Knox Company, 23, 110
Bloomberg, Michael R., xvii, 128
BMT (see Brooklyn-Manhattan Transit Company)
Board of Water Supply, 23
Booth and Finn, 25
Boring machines, 47
Bovis Lend Lease, 125

Bowery Bay Interceptor Sewer and Treatment Plant
    Extension, 87
Bowery Bay sewage treatment plant, 60
Bowery Bay Wastewater Treatment Plant, 8
Bradley Contracting Company, 20, 23
Bradley-Gaffney-Speers, 23
Bridges:
    1909–1929, 24–25
    1946–1962, 78–79
    1963–1979, 97–98
    1980–2009, 123
    timeline of, 6–7
Broadway Line, 37
Bronx, 16, 54
Bronx River Parkway, 12, 78, 131
Bronx-Whitestone Bridge, 7, 81
Brooklyn, 16, 54
Brooklyn Army Terminal, 27
Brooklyn Bridge, 21, 35, 39, 95, 116, 117
Brooklyn Queens Expressway, 128
Brooklyn Rapid Transit (BRT), 37, 42
Brooklyn Union Gas, 29
Brooklyn-Battery Tunnel, 13, 45, 56, 78
Brooklyn-Manhattan Transit Company (BMT), 20,
    42, 116
Brooklyn-Queens Expressway, 57, 78, 79
BRT (see Brooklyn Rapid Transit)
Bruckner Expressway, 79
Brunel, Marc, 44
Bryson, Thomas B., 152
Bureau of Water Supply, 101
Bush, Irving, 27
Bush Terminal complex, 27

C

Cabot & Rollins Corp., 23
Caisson disease, 39
Caissons, 37, 39, 66
Cannonsvile Reservoir, 101
Capital spending, 119
Carey, James L., 153
Carleton Company, 25
Catskill Aqueduct, 23, 24, 42, 43, 47, 68, 109, 110
Catskill reservoir, 84
Cayuga Construction Corp., 43, 78–79, 103
Centre Street, Manhattan, 88, 89

Centre Street Loop, 10
Charles F. Vachris, Inc., 54, 78
Chase Manhattan Bank headquarters, 68–69
Chelsea Piers, 27, 30
Chemical soil stabilization, 69
Chrysler Building, 63
City Hall, xviii
City Planning Commission, 114, 116
City Water Tunnels (see under Water tunnels)
Civetta, Theodore, 72, 96, 156
Civetta Cousins, 125
Civil rights rally, 95
Clearview Expressway, 79
Clove Lakes Expressway, 79
Coal famine (1918), 30
Cochrane, Thomas, 39
Cofferdams, 66
Cold War, 76–77
Collins, Howard A., 155
Columbus Circle, iv, v
Concrete, 108–111
    bulk, 53, 111
    forms for, 110
    mixing, 109
    in road building, 130
Conesco Doka, 125
Coney Island sewage treatment plant, 60, 128
Coney Island Wastewater Treatment Plant, 8
Consolidated Edison (Con Edison), 100
Containerized shipping, 82
Coolidge, Calvin, 25
Corbetta Concrete Corp., 25
Corbetta Construction Co., 101, 123
Correll Contracting Corp., 25
Corona Line, 16, 20
Cortlandt Street Station, 121
Cranford, Frederick L., 152
Cranford Company, 53, 111, 132
Crime, 118–119
Crimmins, Thomas, 152
Cross Island Parkway, 57
Cross-Bay Bridge, 98
Cross-Bronx Expressway, 13, 78
Croton Aqueduct, 23, 41, 124
Croton Filtration Plant, 9, 124
Croton reservoir, 84

Cruz, Edward, 157
Cut and cover tunneling method, 42–43

D
Daily News, 116
Daniel A. Moran, 39
Default on city debt, 96
Degnon Contracting Company, 23, 26
Degnon-McLean Construction Company, 20
Del Balso Construction Corporation, 78
Delaney Associates LP, 128, 129
Delaware Aqueduct, 8, 42, 47, 101
Department of Design and Construction, xv
Department of Marine and Aviation, 82
Department of Sanitation, 58
Department of Street Cleaning, 58
Design Commission, xvii
Dewatering, 65–69, 106
Diebitsch, Emil, 152
Digging, 34–40
  bedrock, 34–36
  below the waterline, 37, 39
  reusing muck from, 39–40
  (See also Tunneling)
DiMenna, Frank P., 155
DiMenna, Nicholas, 156
Dinkins, David, 119
Donohoe, John, 34, 65, 106, 127
Dragados USA, Inc., 114
Dravo Corporation, 23
Drill and blast tunneling method, 47
Drummond, Walter J., 152

E
E. E. Cruz and Company, Inc., 123, 126
Earth Day, 100
East 73rd Street Incinerator, 83
East River Drive, 13, 57, 58, 64, 79, 110, 111
East River Plant (Con Edison), 29
East Side Access Tunnel, 68, 114, 115, 127
Eastchester Bridge, 98
Eisenhower, Dwight D., 77
Electricity (1909–1929), 29
Elevated trains, 81, 90
Empire State Building, 53
Energy conservation, 100

Environmental movement, 98, 100
Excavation (see Digging; Tunneling)

F
FDR Drive (see Franklin D. Roosevelt Drive)
Fehlhaber Pile Co., 79
Ferrara, Joseph, 108, 111
Ferrara Brothers, 125
50th Street Station, 38, 39
Fire engine, horse-drawn, 32
Fire of 1835, 23
Fiscal crisis (1970s), 96
Fitzgerald, F. Scott, 58
Floyd Bennett Field, 56, 60
Flushing Bridge, xviii
Flushing Line, 10, 28, 29
Flushing Meadows Corona Park, 54, 98
Foley Brothers, Inc., 25
Ford, Gerald, 96
Foundation Company, 63–64, 68
Foundations, 62–64
  digging for, 34–40
  innovations for, 103
  underpinning, 88, 90
Fourth Avenue Line to 95th Street, 11
Fox, George A., 156
Franklin Avenue Station, 116
Franklin D. Roosevelt (FDR) Drive, 57, 111, 122, 123
Franklin Shuttle Line, 116
Frederick Snare Corporation, 25, 79
Freedom Tower, 126
Freezing construction sites (dewatering), 66, 69
Fresh Kills Landfill, 83, 84
Fresh Kills Landfill park, 124
Full Freight Access Program, 124

G
Gaffney Gahagan Construction Corp., 26
Gahagan, Walter H., 153
Garber, Daniel A., 152
Gateway, 125
General Contractors Association of New York
  (GCA), xv, 1, 19
  Annual Dinner (1915), 24
  as asset, 128

  during economic depression, 53
  formation letter, 164
  members, 134–151
  memorabilia, 164–165
  presidents, 152–157
George B. Fry, 23
George C. Rogers & Co., Inc., 27
George Fuller Co., 74
George H. Flinn Company, 45
George J. Atwell Foundation Corporation 57
George M. Brewster & Sons, Inc., 25, 68
George W. Rogers Construction Corporation, 83
George Washington Bridge, v, vii, 6, 25, 34, 35
Gilbert, Cass, 25
Gilhooley, John, 3
Giuliani, Rudolph, 119
Goethals, George W., 24
Goethals Bridge, 6, 25
Goodman, William, 156
Gowanus Expressway, 57, 79
Gowanus Parkway, 12
Grace Construction Company, 125
Grand Central Parkway, 57, 61, 121
Grand Central Parkway Extension, 130, 131
Grand Central Station (Terminal), 16–17, 26, 103, 122–123
Granite Construction, 126
Grant Smith Company, 23
Great Depression, 50, 53
Green, Andrew Haswell, 16
Greenpoint/Newton Creek natural gas plant, 29
Ground Zero, 125–126
Grout, 66, 68, 69
Grow Tunneling, 54
Grow-Kiewit-Morrison-Knudsen, 45
Guggenheim Bandshell, 94, 95
Gull, Anthony G., 155
Gull Contracting Co., Inc., 40, 78
Gunn, David, 121

H
H. S. Kerbaugh, Inc., 23, 26
Haberman, Clyde, 96
Haggarty, Harry, 153

Haggerty, Joseph J., 155
Harlem River Drive, 78
Harlem River Rail yards, 124
Hell Gate Bridge, 26
Hell Gate electricity station, 29
Henry Hudson Parkway, 12, 56, 57
Henry Steers, Inc., 26
Herrick, E. A., 153
Hewitt, Abram, 20
Highway Act (1956), 77, 79, 81
Highways:
  1909-1929, 24
  1930-1945, 57
  1946-1962, 78–79
  1963-1979, 98
  1980-2009, 123
  asphalt paving, 130–133
  timeline of, 12–13
Historic buildings/neighborhoods, xvii, 103
Holbrook, 23
Holland, Clifford Milburn, 5, 24
Holland Tunnel, 5, 12, 24–25, 45, 64
Hoover Dam, 108
Horn Construction Company, Inc., 40, 79
Hornstein, Moses, 155
Horrace Harding Boulevard Sewer, 58, 59
Horse-drawn fire engine, 32
Horses, 58
Howland Hook Container Terminal, 119
Howland Hook Marine Terminal, 124, 125
Hudson, Henry, 16
Hudson River, 101
Hudson River Vehicular Tunnel, 64
Hudson River water station, 84, 87
Hudson-Fulton Celebration, 16
Hugo Neu Corporation, 124
Hunts Point market, 103
Hunts Point natural gas plant, 29
Hunts Point sewage treatment plant, 84
Hunts Point Wastewater Treatment Plant, 8
Hutchinson River Parkway, 12, 57
Hutchinson River Parkway Extension, 32, 33, 56
Hylan, John F., 27

I
ICANDA, Ltd., 69

Idlewild Airport, 56, 81, 105
Incinerators, 60, 83, 98
Independent Subway Line (IND), 21, 39, 40, 42, 90
Infrastructure, 1
Interboro Parkway, 57
Interborough Rapid Transit Company (IRT), xviii, 4, 10–11, 16, 20, 38–42
Interstate Commission on the Delaware River Basin, 84
Iovino, Thomas, 94, 115, 127, 128, 157
IRT (see Interborough Rapid Transit)

J
J. F. Cogan Company, 23
J. Rich Steers, Inc., 79
Jacob Javits Convention Center, 114, 119
Jacqueline Kennedy Onassis Reservoir, 23
Jamaica Bay, 27
Jamaica sewage treatment plant, 60
James Stewart & Co., 57
Javits, Jacob, 4, 114
JetBlue, 128, 129
JFK Airport (see John F. Kennedy International Airport)
Job growth, 118
John C. Rodgers Company, 20
John Civetta & Sons, 95
John F. Kennedy International Airport, 56, 104, 105, 114, 121, 128, 129
John Meehan and Son, 25
John P. Picone, Inc., 111, 123, 124
Johnson, Arthur A., 153
Johnson, Drake & Piper, Inc., 78
Johnson-Kiewit, 84
Joseph Miele Contracting Co., 68
Judlau Contracting, 114, 116, 125
Julliard School of Music, 94, 95
Junge, Bruce, 133

K
Karl Koch Erecting Company, 77, 106, 118, 119
Kennedy Airport (see John F. Kennedy International Airport)
Kensico Dam, 24
Kiewit Construction, 126
Kiley, Richard, 121

Kill Van Kull rail bridge, 27
King, Thomas, 41, 43, 128, 157
"The Kiss" photo, 73
Koch, Edward, 118
Koch, Robert, 22, 107
Koch Skanska, 125

L
LaGuardia, Fiorello, xviii, 53, 60, 81
LaGuardia Airport, 56, 61
Landfills, 54, 58, 60, 83, 84
Landmark demolition (1963-1979), 103
Landmarks Preservation Commission, xvii
The Laquilla Group, 126
Lazar, Daniel M., 156
Le Corbusier, 74
League of Nations, 73
Lie, Trygve, 74
Lincoln Center, 94, 95, 98
Lincoln Tunnel, 12, 44, 47, 53, 57
Lindsay, John, 4
LIRR (Long Island Railroad), 121
Locher, 23
Long Island, 54
Long Island Expressway, 12, 57, 79
Long Island parkway system, 53, 57
Long Island Railroad (LIRR), 121
Lower Manhattan Development Corporation, 126

M
Macadam, 130, 131
MacArthur Brothers Company, 23
MacDonald, John S., 154
MacIsaac, Frederick J., 152
MacLean, Mansell L., 155
MacLean-Grove and Co., 57
Madison Square Garden, 103
Maintenance (1946-1962), 76
Major Deegan Expressway, 78
Mancini, Sal, 127
Manhattan, 16, 35, 54, 65, 74
Manhattan Bridge, 6, 23, 56, 81
Manhattan Life Insurance Company Building, 39
Manufacturers Hanover Trust Building, 66

Mason & Hanger Co., Inc., 23, 53, 57
Mass transit:
  1909-1929, 18, 20-21
  1946-1962, 81
  1963-1979, 105
  1980-2009, 121-122
  timeline of, 10-11
  (See also Railroads; Subway system)
McAdam, John, 130
McCarthy, "Fishhook," 58
McClellan, George B., Jr., 23
McDonald, William P., 154
McGovern, Patrick, 153
McMenimen, William V., 153
Meehan, Joseph S., 154
Merguerian, Charles, 24
Merjan, Stanley, 62, 88
Merritt-Chapman & Scott Corp., 79
Metro-North Railroad, 122
Metropolitan Museum of Art, 116
Metropolitan Opera House, 94, 95
Metropolitan Transportation Authority, 122
Mitchell, John Purroy, 5
Moore, Thomas, 68
Moretrench, 66, 68, 106, 125
Moriarty, James, Jr., 17, 69, 88, 96, 122, 128, 157
Moses, Robert, 53, 54, 77-79, 81, 83, 87, 98
Mosholu Parkway, 57
"Muck," moving, 39-40
Municipal Asphalt Plant, 132

N
NAB Construction Corp., 101, 123, 132
Natural gas (1909-1929), 29
Necaro Co., 54
Needle beams, 88
Neumann, Gerard, 156
Neumann, Gerard, Jr., 156
Neversink Reservoir, 87
New Croton Aqueduct, 23
New Croton Dam, 19
New England Thruway, 78, 79
New York Central Harlem River Bridge, 7
New York Central Railroad, 26

New York City Landmarks Preservation Commission, 103
New York Connecting Railroad Bridge, 50, 51
New York Edison Company, 29
New York State Theater, 94, 95
New York World's Fair, 79
Newtown Creek sewage treatment plant, 69, 101
Nicholas DiMenna & Sons, Inc., 42, 56, 58, 62, 101, 123
Nicholson Construction, 125, 126
Nicklas, C. Aubrey, 153
1909-1929, 16-33
  bridges and tunnels, 24-25
  electricity, 29
  guiding growth, 19
  natural gas, 29
  Port of New York, 26-27
  Port of New York Authority, 32
  railroads, 26
  Regional Plan, 32
  subways, 20-21
  telephones, 29
  timeline, 14-15
  water supply, 23-24
  World War I, 300
1930-1945, 50-62
  highways, 57
  LaGuardia Airport, 56
  timeline, 48-49
  Triborough Bridge, 54, 56
  tunnels, 57
  waste management, 58, 60
  World War II, 60-61
  World's Fair, 54
1946-1962, 73-87
  bridges, 78-79
  highways, 78-79
  maintenance, 76
  mass transit, 81
  new building, 77
  new zoning ordinance, 87
  NYC as economic/cultural center, 73-74, 76
  Port Authority, 81-82
  railroads, 81-82
  sanitation, 83-84
  timeline, 70-71
  tunnels, 78-79
  water supply, 84, 87
1963-1979, 95-107
  bridges, 97-98
  environmental movement, 98, 100
  highways, 98
  infrastructure innovations, 103
  landmark demolition, 103
  mass transit, 105
  sewage treatment, 101
  timeline, 92-93
  urban crisis, 95-97
  water supply, 101
  World Trade Center, 105-106
  World's Fair (1964), 98
1980-2009, 114-129
  bridges, 123
  construction increase, 119, 120
  Grand Central Terminal, 122-123
  highways, 123
  mass transit, 121-122
  Port Authority, 125
  rail freight, 124
  recovering from terrorist attack, 125-126
  sanitation, 123-124
  sewage, 123-124
  timeline, 112-113
  urban revival, 114, 116, 118-119
  water supply, 124
North River sewage treatment plant, 101, 123
North River Water Pollution Control Plant, 101
Northern State Parkway, 57

O
Oak Point Line, 124
Old Croton Aqueduct, 23
Onassis, Jacqueline Kennedy, 103
O'Rourke Engineering Construction Company, 26, 90
Outerbridge Crossing, 25, 26
Owen, David, 108
Owl's Head sewage treatment plant, 84, 85, 87

P
Paerdegat Basin, 123

Parsons, William Barclay, 4, 20, 42, 47, 126
PATH trains, 26, 39, 41, 125, 126
Patrick McGovern Company, 20, 39
Paving, 130 (See also Asphalt paving)
Pegno, John, 157
Pelham Parkway, 57
Penn Station (see Pennsylvania Station)
Pennsylvania Plaza complex, 103
Pennsylvania Railroad, 26, 41
Pennsylvania (Penn) Station, 26, 30, 31, 103
Pepacton Reservoir, 87
Perini & Sons, Inc., 101
Perini Corporation, 101, 123
Peter Kiewit Sons Company, 79
Philharmonic Hall, 94, 95
Piers, 82
Piles, 62–63, 103
Pittsburgh Contracting Company, 23
PLAN NYC, 128
Planned shrinkage policy, 114
Planning Commission, xvii
PlaNYC, xvii
Pneumatic process foundations, 39
Poirier & McLane, 40, 56, 57, 69, 78–79, 101, 123
Population, 118
Port Authority of New York and New Jersey, 25–27, 57
    1909–1929, 32
    1946–1962, 81–82
    1980–2009, 125
    during World War I, 30
Port Richmond sewage treatment plant, 84
Post & McCord, Inc., 27, 56
Prospect Expressway, 79
Public schools, 116, 118
Pumps, groundwater, 66, 68

Q
Queens, 16, 54, 56
Queens Midtown Tunnel, xviii, 12, 52, 53, 57
Queens Structure Corporation, 79
Queens trunk sewer, 42

Queensboro Bridge, 6, 18, 97
Queensboro Bridge Plaza, 54, 55
Queensbridge Park Tunnel, 98

R
Rail freight, 124
Railroads:
    1909–1929, 26
    1946–1962, 81–82
    1980–2009, 124
Ransom concrete mixer, 108
Ravitch, Richard, 121
Real estate values, 118
Recycling, 100, 124
Red Hook Marine Terminal, 125
Red Hook sewage treatment plant, 123
Regional Plan Association, 32
Regional Plans, 32
Reid, Hugo, 152
Reilly, John A., 154
Remington, Franklin, 152
Reservoirs, 23
Rice & Ganey, Inc., 23
Riis, Jacob, 16
Rikers Island, 54
Rikers Island Bridge, 98
Riverside Drive Viaduct, 62
Riverside Park, 40
Roads, timeline of, 12–13 (See also Highways)
Rockefeller, Nelson A., 4, 74
Rockefeller Center, 36, 76
Rogers Corporation, 60
Ronan, William J., 4
Rondout Reservoir, 8, 87
Roosevelt, Franklin D., 53, 76
Roosevelt airfield, 60
Roosevelt Island Bridge, 79
Rusciano & Son Corp., 25, 78

S
Sackett, Arthur J., 154
Safe Streets program, 119
Sanitation:
    1930–1945, 58, 60
    1946–1962, 83–84
    1980–2009, 123–124

Saunders, John D., 156
Scannel, Daniel, 3
Schiavone Construction Co., 101, 123–124
Schools, 116, 118
Sea-Land Company, 82
Second Avenue Subway Line, 43, 81, 102, 103, 105, 122
Second Avenue Tunnel, 11
Senior & Palmer, Inc., 25
September 11th terrorist attack, xvii, 125
Seventh Avenue line, 20
Sewage treatment plants, 84
Sewers and sewage, 82
    1930–1945, 60
    1963–1979, 101
    1980–2009, 123–124
    timeline of, 8–9
    tunneling for, 41
Sheridan Expressway, 79
"Shield" tunneling, 44–45
Shore Parkway Bridge, 123
Sicilian Asphalt Paving Company, 20
Silver Lake water storage tanks, 101
Simpson, Edward, 157
Sixth Avenue Subway, 90, 91
63rd Street Tunnel, 11, 36, 96–98, 105, 121
Skanska Koch, 114, 118, 126
Skanska Mechanical and Structural, 127
Skanska USA Civil Northeast, 122–124, 126, 127
Skyscrapers, 19
Slattery Construction Corp., 123
Slattery Contracting Company, 25, 40, 54, 74, 78–79, 81, 90, 101, 103
Smeaton, John, 108
Smith, Alfred E., 53
Snare & Triest Company, 23
Sonic pile-drivers, 103
Spearin, Preston & Burrows, 30, 40, 74, 82
Spencer, Charles B., 154
Spencer, White and Prentis, 40
Spooner, Ray N., 153
Spring Creek Bridge, 57
Springfield Boulevard Sewer, 82
Stapleton piers, 27
Starr, Roger, 114

Staten Island, 16, 54, 79, 83, 97, 124
Staten Island Anchorage, Narrows Bridge, 84
Staten Island piers, 27, 60
Statue of Liberty, 16
Steers, J. Rich, 154
Steers-Snare, 79
Steinway Tunnel, 2
Sternlieb, George, 114
Submerged tunneling tubes, 45–47
Subway system, 3–4, 11, 46, 127
    1909–1929, 20–21
    1946–1962, 81
    1980–2009, 121–122
    Brooklyn-Manhattan Transit Company, 20
    Independent Subway Line, 21, 42, 90
    Interborough Rapid Transit Company, 20–21
    origin of, 20
    and September 11 attack, 126
    underpinning challenges with, 90
Sutton, Percy E., 4

T
T. A. Gillespie Company, 23
T. B. Bryson, 23
T. Moriarty and Son, 69, 96, 122
Tallman Island sewage treatment plant, 58, 60
Talmadge, Richard E., 155
Tarmac, 130
Telephones (1909–1929), 29
Ten Year Plan for Housing, 118
Terminal Construction Corp., 123
Terry & Tench, 27
Third Avenue Bridge, 123
Thomas Crimmins Contracting Company, 23, 43,
    103
Thomas McNally, 23
Throgs Neck Bridge, 7, 79
Timelines:
    1909–1929, 14–15
    1930–1945, 48–49
    1946–1962, 70–71
    1963–1979, 92–93
    1980–2009, 112–113
    bridges, 6–7
    mass transit, 10–11
    roads and tunnels, 12–13

water and sewers, 8–9
Times Square, 73
Trains (see Elevated trains; Railroads; Subway
    system)
Triborough Bridge, xvi, 6, 50, 51, 53–54, 56, 131
Triborough Bridge and Tunnel Authority, 57
Trolley lines, 81
Truman, Harry S., 76
Tully, Edward A., 154
Tully, Gerard P., 156
Tully, Kenneth, 156
Tully, Peter, 51, 130, 157
Tully & DiNapoli, Inc., 40, 54, 56, 78
Tully Construction Company, 124–126
Tunnel Boring Machines, 114
Tunneling, 41–47
    boring machines for, 47, 114
    by boring through bedrock, 44
    cut and cover method, 42–43
    drill and blast method, 47
    timeline of, 12–13
    (See also Water tunnels)
Turner Construction, 74, 125
TWA Terminal at JFK Airport, 104, 105
26th Ward Wastewater Treatment Plant, 8
225th Street Bridge, 81

U
Ulen & Company, 23
Unbekant, Donald, 157
Underpinning, 88, 90
Underpinning and Foundation Company, 20, 60, 64,
    88
Union labor, 128
United Nations, 73, 74
United Nations Building, 74, 80, 81
Urban crisis, 95–97
Urban Mass Transit Administration, 105
Urban revival, 114, 116, 118–119

USS Texas, 30

V
Van Wyck Expressway, 13, 72, 73, 78, 79, 98,
    133
Verrazano-Narrows Bridge, 7, 53, 79, 81, 97
Vivian Beaumont Theater, 94, 95
Volpe, John A., 4

W
Wagner, Robert F., III, 116
Wall Street, reconstruction of, 120, 121
Walsh Construction Co., 57, 74
Ward's Island sewage treatment plant, 60
Ward's Island Wastewater Treatment Plant 8
Warren & Wetmore, 27
Washington Market, 103
Waste management, 58, 60 (See also Sani a
    tion)
Wastewater treatment plants, 8–9
Water conservation, 84
Water pollution, 58, 60
Water supply:
    1909–1929, 18, 23–24
    1946–1962, 84, 87
    1963–1979, 101
    1980–2009, 124
    first, viii
    timeline of, 8–9
    water valve chamber, ix
    (See also Aqueducts)
Waterline/water table:
    dewatering, 65–69
    digging below, 37, 39
Waterside Plant (Con Edison), 29
Waterways, building challenges related to, 35
Watts, John J., 153
Weeks Marine, 125
Welfare Island Tunnel, 96, 97
Wellpoint dewatering method, 68
West Shore Expressway, 79
West Side Highway, 53, 57, 95

Water tunnels, 23
    City Water Tunnel #1, 5, 24, 47, 101
    City Water Tunnel #2, 47, 101
    City Water Tunnel #3, 9, 45, 47, 101, 124

West Side Interceptor Sewer, 101
White, Edward A., 155
White Construction, 118
Whitestone Expressway, 57
Whitestone Expressway Bridge, 123
Wilgus, William J., 24, 45
William J. Fitzgerald, 25
Wilson, Charles W. S., 154
Wilson, Woodrow, 73
Winston & Company, 23
Woman suffrage parade, 19
Woolworth Building, 18, 35, 63

World Trade Center, 105–107
  construction of, 84, 85
  destruction of, 125
  dewatering for, 66, 67, 69
  excavated spoil from, 40
  foundation for, 103
  Ground Zero cleanup, 125–126
World Trade Corporation, 82
World War I, 300
World War II, 60–61, 73, 81
World's Fair (1939), 54
World's Fair (1964), 98

World's Fair Train car, 98, 99
Wright, Wilbur, 16

Y
Yankee Stadium, 21, 118
Yonkers Contracting Co., 79, 125–126

Z
Zeckendorf, William, 74
Zoning ordinances, 19, 87